INTRODUCTION TO
MARKET ACCESS FOR
PHARMACEUTICALS

INTRODUCTION TO
MARKET ACCESS FOR
PHARMACEUTICALS

Mondher Toumi

CRC Press
Taylor & Francis Group
Boca Raton London New York

CRC Press is an imprint of the
Taylor & Francis Group, an **informa** business

CRC Press
Taylor & Francis Group
6000 Broken Sound Parkway NW, Suite 300
Boca Raton, FL 33487-2742

© 2017 by Taylor & Francis Group, LLC
CRC Press is an imprint of Taylor & Francis Group, an Informa business

No claim to original U.S. Government works

Printed on acid-free paper
Version Date: 20161102

International Standard Book Number-13: 978-1-138-03218-7 (Paperback)

Visit the Taylor & Francis Web site at
http://www.taylorandfrancis.com

and the CRC Press Web site at
http://www.crcpress.com

Contents

Contents

Foreword

With this original contribution to health economics, Professor Mondher Toumi has done us all a great service by defining and illustrating the role and utility of the term, and the concept of, *market access*. Make no mistake about it; there are markets for terminology and ideas, and therefore competition among them. In this book, he explains the concept of market access and makes a convincing case for its prevalence and use in health economics and for its popularity among pharmaceutical manufacturers.

The use of market access as an analytical term is relatively a new development in health economics. Based on Google Ngram, the use of the term *health economics* in books came into use after 1970: Its annual use grew dramatically between 1970 and 1978, leveled off during the period of 1978–1983, then grew strongly until 2003, and has leveled off again from 2003 to 2006. Interestingly, the overall Ngram pattern for *market access* is similar but growth is steadier until a recent plateau, and interestingly, it is the more commonly found phrase. But a quick search of PubMed keywords for the published scientific literature finds some 15,000 articles using *health economics* as a keyword versus only about 600 articles using *market access* as a keyword. On the other hand, a simple English-only search for *health economics* yields more than four million Google hits, whereas *market access* yields more than seven million. But "market access" "health" yields only about two million hits. These patterns will make sense as I recount below Professor Toumi's explanation of the origin of the term market access.

Health economics emerged as a subfield of economics in the early 1970s, when university-level courses began to be taught. Professor Toumi cites Professor Kenneth Arrow's seminal 1963 article on "Uncertainty and the Welfare Economics of Health Care." It identified the unique characteristics of medical care and health as economic goods. Owing to uncertainty of various kinds, these markets can fail to perform optimally, that is, achieve Pareto efficiency, and many of the regulatory interventions that are evident can be explained as an evolutionary response—both from the public and private sectors—to these market failures.

Professor Toumi is also a leading thinker in the subfield of health economics called *pharmacoeconomics*. We have also witnessed the tension of terminology within this subfield. Taking an economics perspective, one could argue that there is indeed a demand and supply for new terms, concepts, and ideas. In addition, it can be a bit of a battlefield: Consider, for example, the general concept of *personalized medicine*, which has variously been called *stratified medicine*, *tailored treatment*, *individualized therapy*, and, mostly recently, *precision medicine*. Proponents of each would argue that these are imperfect synonyms and that each has some unique aspect that is valuable. These words can have power and influence as well: President Barack Obama launched a "Precision Medicine Initiative" in January 2015, reflecting a belief that the corresponding clinical science has evolved to point where genetic and other biomarkers can predict individual response to therapy. Manufacturers, academics, consultants, and even government all compete in this market to differentiate the value that their concepts can provide.

Most readers of this book and even most health economists will not be familiar with the original use of the term *market access*, as explained well by Professor Toumi, in the field of international trade by the World Trade Organization. Clearly, thinking internationally, that is, with nation states competing to serve their own citizens first, world economic history has been replete with all sorts of barriers—both tariff and nontariff—to trade. He defines market access as "… the ability for a drug to achieve through a health insurance system a reimbursed price and a favorable recommendation for prescription." For a manufacturer interested in selling its products in as many national markets as possible, the term *market access* means overcoming those barriers to get one's product on the market in a particular country. For innovative pharmaceuticals, the substantial barriers are market authorization, price and reimbursement negotiations, and controls on utilization. Of course, the concept can also apply to specific insurance market segments in a complex and fragmented public and private insurance market such as the one that the United States has.

Professor Toumi and his colleagues have surveyed pharmaceutical companies and found that they organize their activities in this area in a variety of ways. Most staff working in this area were originally trained in fields such as decision sciences, pharmaceutical administration, health economics, epidemiology, biostatistics, psychometrics, clinical pharmacology, and business administration. But with the complementarity of these fields being fairly obvious, especially when confronted with a range of regulatory, health technology assessments, and pricing and reimbursement barriers, companies have often grouped some of these activities, particularly at a global level, under the umbrella of *market access*. Having worked in a global health economics group in a large company, and having taught in a PhD program to train interested researchers with the capability to work in these company units, I have seen a gradual evolution in the recognition that staff working in these groups need basic training in all of these areas and an increased understanding of the market access barriers in these markets: so our curriculum continues to evolve in this direction. To my knowledge, Professor Toumi is the pioneer here—the first to create an academic department that explicitly offers degrees in market access.

More than 200 years ago, Adam Smith pointed out in his book *The Wealth of Nations* that "division of labor is limited by the extent of the market." Professor Toumi is probably ahead of the curve here, but it is important to recognize that academic markets are competitive and will be somewhat resistant to change, even if it only means renaming the integrated, multidisciplinary training they are already doing under a new label. It is difficult to think of a better term than *market access* that captures the essence of these activities. Even in my own 12 years working in the industry, our unit evolved from *pharmacoeconomics* to *health economics* (to consider diagnostics and devices), and to *economics* (to add pricing and reimbursement considerations). And subsequently this has evolved to *payer strategy* (to focus on strategic thinking to overcome market access barriers). The conceptual innovation of grouping these activities under *market access* is understandably appealing, and it is essentially facing a market test at the moment.

I have considerable doubt about whether and when this might reach a tipping point resulting in market access becoming the most widely used term for these activities. But I have no doubt that Professor Toumi has done all of us a great service by exploring and clarifying the utility of this concept, and identifying its relevance to a wide range of phenomena such as health technology assessment, market access agreements, value propositions, and value-based pricing.

Although I have emphasized the novelty of Professor Toumi's case for the unifying concept of market access, readers of this book will also be astounded by the breadth and depth of his knowledge and understanding of the real-world institutions affecting market access in each of the individual chapters for nine major pharmaceutical markets: France, Germany, Italy, Spain, Sweden, the United Kingdom, Belgium, the United States, Japan, and China. In addition, these themes are extended to the related specific issues, such as the market for vaccines, orphan drugs, external reference pricing, early advice, early access, and market access agreements. I would encourage you to read this book, and learn and enjoy contemplating these important insights and connections that Professor Toumi is sharing with us.

Lou Garrison
Pharmaceutical Outcomes Research & Policy Program
Department of Pharmacy
Departments of Global Health and Health Services
University of Washington

Acknowledgments

I thank everyone who helped me complete this book by providing review, input, comments, and continuous encouragement. Dear all, without your effective support, I would have not been able to bring my work to a successful completion.

Very special thanks to Szymon Jarosławski and Cécile Rémuzat for your continuous review and input along this book's development.

Thanks to Samuel Aballéa, Lieven Annemans, Charles Sabourin, Anna D'Ausilio, Emna El Hammi, Jaime Espin, Clément François, Shunya Ikeda, Patrycja Jaros, Åsa Kornfeld, Stefan Lhachimi, Abayomi Odeyemi, Florence Baron Papillon, Steve Sherman, Marion Thivolet, Keith Tolley, and Katherine Eve Young for your review.

Thanks to Pr. Lou Garrison and Lieven Annemans for your preface and postface.

To my family for their continous support!

Un clin d'œil à la Commune de Paris

J'aimerai toujours le temps des cerises
C'est de ce temps-là que je garde au coeur,
Une plaie ouverte.
Et Dame Fortune en m'étant offerte
Ne pourra jamais calmer ma douleur...
J'amerai toujours le temps des cerises
Et le souvenir que je garde au coeur.

Le temps des Cerises de Jean Baptiste Clément

Author

Mondher Toumi, MD, MSc (biostatistics), MSc (biological sciences, option pharmacology), PhD (economic sciences) is a professor of public health at Aix-Marseille University, Marseille, France. After working for 12 years as research manager in the Department of Pharmacology at the University of Marseille, he joined the Public Health Department in 1993. From 1995, he pursued a career in the pharmaceutical industry for 13 years.

Mondher Toumi was appointed global vice president at Lundbeck A/S in charge of health economics, outcome research, pricing, market access, epidemiology, risk management, governmental affairs, and competitive intelligence. In 2008, he founded Creativ-Ceutical, an international consulting firm dedicated to support health industries and authorities in strategic decision-making. In February 2009, he was appointed professor in the Department of Decision Sciences and Health Policies at Claude Bernard University Lyon 1, Lyon, France. In the same year, he was appointed director of the chair of public health and market access. He launched the first European University Diploma of Market Access (EMAUD), an international course already followed by approximately 400 students. In addition, he recently created the Market Access Society to promote education, research, and scientific activities at the interface of market access, health technology assessment (HTA), public health, and health economic assessment. He is the editor-in-chief of the *Journal of Market Access and Health Policy (JMAHP)*, which was recently granted PubMed indexation.

Toumi is also a visiting professor at Beijing University (Third Hospital), China. He is a recognized expert in health economics and an authority on market access and risk management. He has authored more than 250 scientific publications and communications, and has contributed to several books.

Common Abbreviations

ACIP	Advisory Committee on Immunization Practices
AEMPS	Agencia Española del Medicamentos y Productos Sanitarios/The Spanish Agency for Medicines and Health
AETS	Agencia de Evaluacion de Tecnologias Sanitarias/The National Health Technology Assessment Agency
AFR	Annual financial return
AGES	Austrian Agency for Health and Food Safety
AHRQ	Agency for Healthcare Research and Quality
AIFA	Agenzia Italiana del Farmaco/Italian Medicines Agency
AMNOG	Arzneimittelmarktneuordnungsgesetz/The Act of the Reform of the Market for Medicinal Products
AMP	Average manufacturer price
ASL	Azienda Sanitaria Locale/Local Health Unit
ASMR	Amélioration du Service Médical Rendu (improvement in actual benefit)
ASNM	Agence nationale de sécurité du médicament et des produits de santé
ASP	Average sale price
AWP	Average wholesale price
BfArM	Bundesinstitut für Arzneimittel und Medizinprodukte/The Federal Institute for Drugs and Medical Devices
BLA	Biologics License Application
BMG	Bundesministerium für Gesundheit/The Federal Ministry of Health
CA	Commercial agreement
CBER	Center for Biologics Evaluation and Research
CDC	Centers for Disease Control and Prevention
CE	Cost-effectiveness
CED	Coverage with evidence development
CEPS	Economic Committee of Health Products
CHIP	Children's Health Insurance Programme
CIPM	Comisión Interministerial de Precios de los Medicamentos/The Interministerial Commission for Pricing of Medicinal Products
CISNS	Consejo Interterritorial del Sistema Nacional de Salud/Spanish Interterritorial Council
CMS	Centers for Medicare and Medicaid Services
CPR	Comitato Comitato Prezzi e Rimborso/Pricing and Reimbursement Committee
CST	Comitato Scientifico e Tecnico/Scientific-Technical Commission
CT	Commission de la Transparence
CTC	Conditional treatment continuation
CTV	Technical Committee on Vaccinations
CVZ	College voor Zorgverzekeringen/The National Health Care Institute
DAHTA	Deutsche Agentur für Health Technology Assessment/The German Agency of Health Technology Assessment
DCE	Discrete choice experiment
DDD	Daily defined dose
DERP	Drug effectiveness review project
DGFPS	Dirección General de Farmacia y Productos Sanitarios/The Directorate General of Pharmacy and Health Products
DH	Department of Health
DIMDI	Deutsches Institut für Medizinische Dokumentation und Information/The German Institute of Medical Documentation and Information
DMA	Danish Health and Medicines Authority
DPO	Drug Pricing Organization
DRG	Disease-related group
DT	Drug tariff
EBA	Early benefit assessment
EMA	European Medicines Agency
EU	European Union
FDA	Food and Drug Administration
FUL	Federal upper limit
G-BA	Gemeinsamer Bundesausschuss/The Federal Joint Committee
GBS	Guillain–Barré syndrome
GKV-SV	Gesetzliche Krankenversicherung-Spitzenverband/The National Association of Statutory Health Insurance Funds
GU	Gazzetta Ufficiale/The Official Journal of the Italian Republic
HAS	Haute Autorité de Santé/French National Authority for Health
HC	Health Council
HEOR	Health Economics and Outcomes Research
HHS	Department of Health and Human Services
HPV	Human papilloma virus
HTA	Health technology assessment
ICER	Incremental cost-effectiveness ratio
INN	International nonproprietary name
IQWiG	Institut für Qualität und Wirtschaftlichkeit im Gesundheitswese/The Institute for Quality and Efficiency in Healthcare
ITR	Index Thérapeutique Relatif
JCVI	Joint Committee on Vaccination and Immunisation
LFN	*Läkemedelsförmånsnämnden*/Pharmaceutical Benefits Board
MA	Market access
MAA	Market access agreement
MAC	Maximum allowable cost
MAH	Marketing Authorization Holder
MAu	Marketing authorization
MCDA	Multicriteria decision analysis
MHLW	Ministry of Health, Labour and Welfare
MHRA	Medicines and Healthcare Products Regulatory Agency
MoH	Ministry of Health

MOHRSS	The Ministry of Human Resources and Social Security
MPA	*Läkemedelsverket*/The Medical Products Agency
MS	Multiple sclerosis
MSSSI	Ministerio de Sanidad, Servicios Sociales e Igualdad/The Ministry of Health, Social Services and Equality
MTA	Multiple technology assessment
NDRC	The National Development and Reform Commission
NHFPC	The National Health and Family Planning Commission
NHI	National Health Insurance
NHS	National Health Service
NICE	National Institute for Health and Care Excellence
NIP	National Immunization Program
NITAG	National Immunization Technical Advisory Group
NVC	National Vaccination Calendar
OECD	The Organisation for Economic Co-operation and Development
OsMED	Osservatorio Nazionale sull'impiego dei Medicinali/National Observatory on the Use of Pharmaceuticals
OTC	Over-the-counter
P&R	Pricing and reimbursement
P4P	Payment for performance
PAS	Patient access scheme
PBAC	Pharmaceutical Benefits Advisory Committee
PBMSHG	Pharmacy Benefits Management Strategic Healthcare
PEC	Department of Defense Pharmacoeconomic Center
PEI	Paul Ehrlich Institute
PFN	Prontuario Farmaceutico Nazionale/The National Pharmaceutical Formulary
POM	Prescription-only medicines
PPRS	Pharmaceutical Price Regulation Scheme
PV	Preventive vaccine
QALY	Quality adjusted life year
RCT	Randomized Controlled Trial
R&D	Research and development
RP	Reference pricing
RSA	Risk sharing agreement
SBU	*Statens beredning för medicinsk och social utvärdering*/Swedish Agency for Health Technology Assessment and Assessment of Social Services
SGB	Sozialgesetzbuch/ Social Code Book
SGCMPS	Subdirección General de Calidad de Medicamentos y Productos Sanitarios/The General Subdirectorate of Quality of Medicines and Health Products
SHI	Statutory Health Insurance
SMC	Scottish Medicines Consortium
SMR	Service Médical Rendu/Actual Medical Benefit
SNS	Sistema Nacional de la Salud/The National Health System
SOP	(Farmaci) Senza Obbligo di Prescrizione/ Nonprescription pharmaceuticals with advertising prohibition
STA	Single technology assessment
STIKO	Standing Vaccination Committee (Ständigen Impfkommission)
TC	Transparency Committee
TFR	Tarif forfaitaire de responsabilité
TLV	*Tandvårds- och läkemedelsförmånsverket*/The Dental and Pharmaceutical Benefits Agency
TVs	Therapeutic vaccines
UNCAM	Union nationale des caisses d'assurance maladie
USP	United States Pharmacopeia
VA	Veterans Affairs
VBP	Value based pricing
VENICE	Vaccine European New Integrated Collaboration Effort
WAC	Wholesale acquisition cost
ZIN	*Zorginstituut Nederland*/National Healthcare Institute

Introduction

Inadequate market access has become the primary reason for commercial failure of any new pharmaceutical launch. It is often referred to as the *fourth hurdle* in reaching the patients, following the need to evidence the traditional three hurdles of efficacy, safety, and quality in order to obtain the regulatory marketing authorization.

Nowadays, market access has become the primary driver for global income of any new drug in development. Without a convincing strategy in place to evidence and communicate the product value to decision makers, the drug will fail to reach its intended commercial value. This requires understanding the key decision drivers.

Market access is a complex process with as many definitions as individuals. Even prestigious publications may contain definitions that are questionable, inappropriate, or inaccurate. This book attempts to harmonize and unify definitions, concepts, and history related to the market access field. The approach presented in this book is situated at a crossroads of multiple disciplines that form an integral part of a successful approach to market access.

This book is based on an accredited course in this area out of the European Market Access University Diploma (EMAUD) affiliated with the Market Access Society, France, and the Aix-Marseille University, Marseille, France.

OBJECTIVES OF THE BOOK

- Present an advanced understanding of the market access environment, concepts, and principles
- Draw a mapping of market access stakeholders
- Present basic knowledge in decision sciences applied to public health
- Review the latest regulations and guidelines in main global markets, including Europe, United States, and Asia
- Review novel pharmaceutical policies such as risk-sharing, managed entry agreements, early entry, early dialogue, vaccine, and orphan drugs market access policies
- Discuss the best practices in market access across case studies

KEY LEARNINGS

- Being able to understand the grounds for a market access strategy with a multidisciplinary approach
- Identify the stakeholders and develop a market access plan that matches their expectations
- Understand specificities of some geographies and some types of pharmaceutical products
- Anticipate the future paradigm changes in market access

WHO SHOULD READ?

This book is intended for students and for industry and authority professionals in the fields of pharmaceuticals, medical biotechnology, and public health.

1 Health as a Good

1.1 WELFARE ECONOMICS AND HEALTH

The health care market has been expanding globally for the past few decades. The demand of the population for better health is increasing steadily, and to face this challenge, *health producers* have to adapt and include economic perspective as a key component of the health care environment.

But can health be considered from an economic point of view? The foundations of the French medicine, for example, are based on the obligation of the practitioners to involve each and every resource necessary for the well-being of a patient.[1] Is this concept really compatible with an economic control of the health consumption? One can strive to achieve maximization of the utility of a population, but is it consistent with minimization of expenditures that needs to be taken into account due to economic constraints?

If we define health consumption as a set of health care products that can be used by the population, then several notions have to be determined in order to understand the specifics of *health goods*, compared to other goods and services of an optimal market as defined by Pareto (1906).[2]

Pareto defined the optimality of a market as an economic state where resources are allocated in such a way that it is impossible to make any one individual better off without making at least one individual worse off.

This implies the following two theorems of welfare economics[3]:

- Any competitive equilibrium leads to a Pareto efficient allocation of resources.
- Any Pareto efficient allocation can be achieved through competition given an appropriate initial allocation of resources.

We would like to introduce a few elementary notions concerning health and health care in order to make sure that terms and concepts will be understood properly.

Health itself cannot be bought, what actually is purchasable is the health care as a proxy of health. As Grossman introduced in 1972, buying health care will allow an individual to invest in his *health capital* that will decrease over time.[4] Thus, the buyer's concern will be the amount of health production which their investment will contribute to their health capital, and it is important to note that *this capital can neither be shared nor traded with others*. Indeed, it is possible to give advice on consequences of risky behaviors or dispense health care, but one cannot pretend to give a part of his/her health directly to another.

In Sections 1.2 through 1.6, we will demonstrate the particularities of the health care market and why this market cannot tend by itself to optimality/efficiency as described by Pareto in 1906.

1.2 HEALTH CARE: A MIXED AND COLLECTIVE GOOD

The health care products cannot be considered like any other goods because, first of all, they have the particularity to be *mixed and collective*. Indeed, they cannot be defined as rival or excludable (like private goods are), but some congestive effects can appear: patients can consume the same medical goods or services at the same time (*nonrival*), nonpaying consumers can have an access to it (*nonexcludable*), but if too many patients are using the same health care product, its production might not be sufficient to be consumed by every patient (*congestive*).

For example, the introduction of a preventive strategy to avoid the diffusion of an infectious disease (such as a vaccine or an educational program) concerns the patient at an individual level (his protection), but also as a part of a community. In the vaccine example, different patients could be able to receive it and people who have not got the injection will still get "protected" by the others, but the number of available vaccines can be restricted and so not everyone would be able to get an injection.

There is also the possibility for those particular goods to induce *externalities*. Externalities are a consequence of an activity on a party who have not decided to generate this activity.[5] A good example of a positive externality would be the benefit of a healthy worker on the results of a company.

Health care can also be considered as public good, because, just like education or public transport, it is a part of the public services provided (at a certain level depending on the country) by the state. Potential consequences of the public aspect could be an equal access to the continuity of care as well as, for certain countries, a universal access.

Key points:

1. Health care is a mixed and collective good
2. Health care can induce externalities
3. Consequence of a health care as a public good: universal access and continuity of care

1

1.3   EQUITY, HEALTH, AND HEALTH CARE

The fact that health is essential to human life creates a need to obtain or sustain the health irrespective of the resources involved. However, people do not have the same resources to invest in their health capital (time or money), which makes them *unequal in front of the health risk and need*. This situation is unacceptable in a fair society. This is why mechanisms have been implemented to ensure that everyone is able to receive the necessary health care for their condition, such as programs that provide insurance for the lowest income population or services that are free of charge at the point-of-use (emergency and some social services).

Key points:

1. Individuals are unequal in front of the health risk and need
2. Mechanisms have been introduced to insure equity (specific insurances and free services at the point-of-use)

1.4   UNCERTAINTY RELATED TO THE DEMAND AND RESULTS

One particularity of the health care as a good is the fact that the demand for health care is not stable throughout time and *it is unpredictable* (Arrow 1963). Indeed, the uncertainty related to the *occurrence* or the *gravity* of an event disturbs the organization of the health care, with potential terrible consequences, in terms of human and economic losses. To illustrate this, we would consider the example of Ebola virus outbreak in 2014 that caused a massive human loss in Africa, but also required a substantial budget to fight against the epidemic.

It is noteworthy that there is often an uncertainty around the expected outcome when using health care. For example, for the treatment of cancer, there is always a risk that a patient will not recover from sickness and also the diagnostic tests are not 100 percent specific or sensitive.

These two aspects of the uncertainty tend to demonstrate that goods and services related to health are *risky* and could be incredibly *costly*.

Key points:

1. Occurrence or gravity of a disease is unpredictable
2. Uncertainty surrounds the expected health production related to health care
3. Goods and services related to health are risky and costly

1.5   PHYSICIAN'S EXPECTED BEHAVIOR

Another source of uncertainty concerns the relation between physicians and the patients. The patients are not fully able to judge the quality and quantity of a health care good or a service. In contrast to other products such as food, consumers are less able to assess the results from the consumption of a medical good or service and to learn from their experiences. Thus, the medical knowledge can be considered a key component of the relationship between physicians and patients: the larger the degree of *information asymmetry* persists in the relationship, the more uncertain the patients are about the outcome of the health care they receive.

This leads the health care good to be considered as a *trust good*: the quality of the product is never revealed to the patient and the perception of the good by the patient will only reflect on the trust related to the physician. Health care goods can also be considered as an *experience good*: when it is difficult to estimate the quality or price of a product, then the experience gained through the consumption will help to better estimate those parameters. It reduces in part the uncertainty about the quality or price, and the reputation or opinion about the physician or a health care product will play an important role in the patient's future decisions.

This biased relationship could induce some discrepancies in physician's behavior. As patients are not able to evaluate the precise accuracy of the physician's actions, the primary objective of the physicians (well-being of the patients, altruism, etc.) could also be altered by their *personal motivation*. This could significantly impact the consumption of goods.

However, the availability of medical knowledge tends to change this information asymmetry and to impact the patient's behavior, who sometimes can develop a wider knowledge on specific topics and so, redefine the relationship with the physician.

Key points:

1. Relationship between health care givers and patients is biased by the information asymmetry related to the quality and quantity of care
2. This relation can also be biased by physician's personal motivation
3. Health care goods are considered as *trust goods* or *experience goods*

1.6   SUPPLY CONDITION

The particularity of the medical goods in terms of supply condition lies in the limitation of the offer through *licensing restrictions*. Indeed, one has to be licensed to offer health care services. The cost of medical education (in some countries), as well as the time and effort spent studying, limits the number of practitioners. However, this allows reducing the risk related to the uncertainty in the patient–physician relationship. Indeed, the education will contribute to ethic, deontology, and quality of care, and so could improve the effectiveness of health care.

The limitation of the offer is also impacted by *the high cost related to* the installation of physicians and the available quality already existing on a competitive market.

Key points:

1. Offers are limited by licensing restrictions and the high cost of the installation
2. Offers are also limited by the number of practitioners related to the cost of medical education and the technicality of the study

1.7 DISCUSSION

These specific characteristics clearly demonstrate that medical goods and services compose a nonoptimal market without an efficient allocation of resources. The market is not regulated by the demand and supply, and the macroeconomic equilibrium is achieved through three elementary components: the demand of care, the supply of care, and the financing. Thus, this market needs to be coordinated by a public or a nonprofit institution.

The uncertainty related to the occurrence or the severity of an event induces a risk that needs to be insured. However, if left entirely to private institutions, the insurers could develop a picking strategy to cover only the healthy members of the society or those at a minimal risk in order to generate profit, and so the universal access to care could be *altered*. Therefore, an intervention of public health authorities is primordial. This is where the regulation has to stand, and the development of social insurance and universal access to health care are its results.

However, even if payers (public or private) reimburse the cost of health care at a know price, there still remains uncertainty concerning the health production associated with this care. Therefore, if the public institutions intervene to control the market, the information asymmetry between the payers (public or private insurers) and the health care *producers* has to be reduced. This can be achieved by mechanisms that control the activity and the quality of care and cost. This has been made, for example, through agreements between insurers, health producers, and external scientific authorities.

However, the control of the quality tends to limit the offer, and the more the offer is restricted, the more the prices are increasing. In addition, the technical innovation in health care that results in novel and costly products and services is putting an ever growing strain on health financing. Also, an ageing population has increasing health needs that can only be met by increasing the expenditure on health care. This dynamic interferes with an economic environment where resources are limited, imposing prioritization of spending. This explains why the discipline of market access for health care goods plays a key role in the health care scene nowadays.

Key points:

1. Medical goods and services compose a nonoptimal market that is not regulated by demand and supply: demand of care = supply of care = funding
2. The intervention of the state is needed for medical goods and services
3. Regulation plays a key role in the health care system, that is, it reduces the inherent uncertainty
4. Growing demand and limited resources for health care result in the need of prioritization to which market access is an answer

REFERENCES

1. Dr. Paul Cibrie. *Charte de la médecine libérale* 1927. Le Congres des Syndicats Medicaux, Paris.
2. Pareto V. *Manuel d'économie politique* 1906. 2nd edn (1927). Translated by A.S. Schwier as Manual of Political Economy. New York: Augustus M. Kelley, 1971. (Citation found in The New Palgrave Dictionary of Economics and the Law by Newman in 1998).
3. Arrow KJ. Uncertainty and the welfare economics of health care. *The American Economic Review* 1963;53:941–973.
4. Grossman M. On the concept of health as a capital and the demand for health. *Journal of Political Economy* 1972;80(2):223–255.
5. Buchanan J. and Stubblebine C. Externality. *Economica* 1962;29(116):371–384. doi:10.2307/2551386.

2 Decision-Making in Public Health

2.1 PUBLIC HEALTH DEFINITION

Public health is a set of private or public measures organized to prevent disease, prolong life, and promote the health of the population. It has a population focus and aims to provide healthy conditions for the whole population.[1] Public health is a complex domain; it is not only concerned with eradicating individual diseases but also with the complete health system.

The health of individuals and communities is affected by many factors. The determinants of health include the following[2]:

- *Social environment*: Income and social status—higher income can be linked to a better health.
- *Education*: Low education can lead to poorer health.
- *Physical environment*: Water, air, workplaces, houses, and roads can affect the health. In fact, indoor pollution, smoking, second-hand smoking, and poor ventilation can lead to serious health problems.
- *Support networks*: Families, friends, communities, and so on. Greater support can enhance individual's health.
- *Gender*: It can play a role because men and women can be affected by different types of diseases.
- *Personal behavior*: Sports, smoking, drinking alcohol, balanced diet, and so on.

2.1.1 EXAMPLE OF THE INFLUENCE OF AIR POLLUTION ON PUBLIC HEALTH

Air pollution is a serious problem caused by several factors such as transportation, industries, and burning of fossil fuels. Poor air quality directly affects human health, principally respiratory and cardiovascular system, and may cause long-term diseases. The health effects caused by air pollutants may range from difficulty in breathing, wheezing, and coughing to serious aggravation of respiratory conditions and lung cancer.

Some decisions should be made to reduce air pollution and avoid its impact on health; people have to be encouraged to change their mode of transport: to use the nonpolluting cars and public transport rather than diesel cars or to walk or use their bikes to reduce pollution, and improve the physical activity.

More importantly, these changes can have an impact on several sectors. For example, they may decrease the turnover volume of car industries and decrease the state income from fuel taxes, but can increase the development of public transport industry in long term.

In addition, air pollution has no frontiers. It is an environmental health concern that has no boundaries. Pollution emitted in one country can have an impact on countries that are halfway around the world. Management decisions to reduce and control air pollution cannot be taken in one country; it must implicate a synchronization between the neighboring countries and beyond.

So making decisions in public health is complex, implicating not only public health sector but also many other sectors. It is a multi- and interdisciplinary decision. An international collaboration is important because some diseases do not obey national boundaries.

2.2 DECISION-MAKING IN PUBLIC HEALTH

2.2.1 ORGANIZATION OF PUBLIC HEALTH

There are several key players: decision makers and payers, responsible for delivering health services to the population. The organization of public health may differ between countries. In general, the ministry of health plays the most important role at the national level. Other health and non-health organizations may deliver some public health services as part of their usual business (Table 2.1).[3]

For example, in the Netherlands, the Ministry of Health, Welfare, and Sport is responsible for the decisions at the national level, the municipalities are responsible at a local level, and there are supporting agencies such as the Council for Public Health, the National Institute for Health Promotion and Disease Prevention, and the National Institute for Public Health and Environmental Issues (RIVM).

In Germany, the Federal Ministry of Health has a small role. The decisions about public health services are mainly made by the federal states (Länder), supported by some agencies such as Federal Institute for Pharmaceuticals and Medical Products, the Institute for Communicable and Noncommunicable Diseases (the Robert Koch Institute), the Federal Centre for Health Education, and the German Institute for Medical Documentation and Information. The Advisory Council for Concerted Action in Health Care provides guidance on monitoring health and economic trends and health care reform and prepares reports on addressing issues such as how to incorporate prevention and health promotion within the social health insurance system.[3]

2.2.2 FUNDING HEALTH CARE AND PUBLIC HEALTH

Different types of health care funding exist[4]:

1. General taxation (the United Kingdom)
2. Local taxation with local councils managing providers (Denmark)

TABLE 2.1

Organization of Public Health in Some European Countries

Organization	National Level	Local Level	Supporting Agencies
Sweden	Ministry of Health and Welfare	The county councils (independent regional government bodies)	National Institute for Public Health National Public Health Committee Commission on National Targets National Board of Health and Social Welfare
Finland	Ministry of Social Affairs and Health (MOSAH)	Municipal health committees	National Public Health Institute Finnish Centre for Health Promotion Intersectoral National Public Health Committee National Research and Development Centre (STAKES)
Denmark	Ministry of Health	Municipalities	The National Board of Public Health (2001) Institute for Clinical Epidemiology The Council on Health Promotion Policy The Danish Council on Smoking and Health
The Netherlands	Ministry of Health, Welfare and Sport	Municipalities	Council for Public Health National Institute for Health Promotion and Disease Prevention National Institute for Public Health and Environmental Issues (RIVM)
France	Multiple governmental bodies	Public health programs in the regions	National Surveillance Agency National Institute for Health and Medical Research High Committee of Public Health and National Institute for Prevention and Health Education Pharmaceuticals Agency French Blood Agency French Committee for Health Education
Germany	Federal Ministry of Health has a small, but growing role	The federal states (Länder) are mainly responsible for public health services	Federal Institute for Pharmaceuticals and Medical Products Institute for Communicable and Noncommunicable Diseases Federal Centre for Health Education German Institute for Medical Documentation and Information Advisory Council for Concerted Action in Health Care

Source: Allin, S. et al., *Making Decisions on Public Health: A Review of Eight Countries*, European Observatory on Health Systems and Policies, London, 2004.

3. Social health insurance paid by the employer and employee, with multiple, noncompetitive, autonomous, and third-party payers (insurers) (France)
4. Social health insurance paid by the employer and the employee, with autonomous, competitive third-party payers (insurers) (Germany)
5. Compulsory social health insurance for basic care paid by individuals with competitive third-party payers (insurers) and government-defined benefit package (Switzerland)
6. Voluntary health insurance predominantly paid by employers with tax subsidies for employers and employees (the United States)
7. Compulsory social health insurance for catastrophic illness and long-term care and social health insurance for acute medical services paid by the employer and employee (the Netherlands)

2.2.3 CRITERIA USED FOR PRIORITY-SETTING AND DECISION-MAKING

Priority-setting is a challenge at global, national, and local level and for all contexts in health systems (Table 2.2).[3]

More details about the decision-making pathways for selected countries are presented in Chapters 12–21 of this book. They generally consist of the following two approaches[5]:

1. *Technical analyses*: Rely on epidemiologic, clinical, financial, or other data. Technical approaches depend on the availability of data, and priorities tend to be based on measurable units such as diseases (burden of disease) and interventions (with respect to their costs and use).
2. *Interpretive assessments*: Rely on consensus views of informed participants.

A decision maker can rely on either of these or on both to a varying extent. For example, in France priorities are based on the burden of illness, societal values and priorities, evidence of inequalities in health outcomes for the health condition or problem within the country, and the current state of knowledge about the condition or health problem's etiology. Technical health economic evaluations are considered in the context of the pricing negotiations held between the manufacturer and the Economic Committee of Health Products (CEPS).[6] Efficiency analyses are considered supplementary data in the

TABLE 2.2

Criteria for Decision-Making in Public Health in Some European Countries

	Criteria for Decision-Making
France	• Significance in terms of burden of illness • Fit with societal values and priorities • Evidence of inequalities in health outcomes for the condition/problem within the country, or poor outcomes in France compared to other countries • Current state of knowledge about the condition/health problem's etiology
Germany	Since priority-setting at the national level is very limited, it is not clear whether a formal mechanism is in place. Responsibilities for decision-making are generally shared between the Länder and the federal government. The process for deciding priority areas in health involves many players, for example, the 2003 National Health Target Document called Gesundheitsziele.de was drafted by more than 70 stakeholder groups and over 200 experts. The methodology consists of reviewing priority-setting criteria in other countries, developing a matrix of morbidity and mortality information on the risk factors/disease areas, and subjecting a short list to political negotiations.
The Netherlands	• Strengthening public health infrastructure • Reduce health inequalities • Encourage healthy lifestyles • Ensure best possible health opportunities for all residents • Promotion of healthy living • Cooperation between the curative care sector and the public health sector • Fostering a coherent policy on public health care, both nationally and locally • Enhancing the administrative and policymaking power of local authorities and municipal health services
Sweden	• Epidemiological, demographic, and household survey data form the basis for local policy-making • Reports published by the National Board of Public Health and Social Welfare informs central policy • Priorities are set based on an *ethical platform* of human dignity, need and solidarity, and cost-effectiveness[20]
Finland	• Local needs assessments and informal reviews of existing programs guide decision-making • Decision-making and priority-setting are inclusive and intersectoral at the local and the national level
Denmark	Decisions are based on evidence of the burden of disease and scope for prevention Main priorities are as follows: • Increase life expectancy and quality of life • Improve equity in health • Strategies to achieve the goals • Legislation (e.g., ban tobacco commercials, ban sales of alcohol to minors) • Information campaigns (e.g., nutritional advice, promoting exercise, and safe sex)

Source: Allin, S. et al., *Making Decisions on Public Health: A Review of Eight Countries*, European Observatory on Health Systems and Policies, London, 2004.

decision-making, alongside the budget impact of the technology, its additional therapeutic value, or other criteria such as industrial matters and support to innovation. In contrast, in the United Kingdom, economic evaluations appear to have a major influence in the evaluation and decision-making process and can predict the large majority of National Institute for Health and Care Excellence's (NICE's) decisions. The question NICE is mandated to answer is "should a technology be employed within the established limited National Health Service (NHS) budget?" NICE does not statute as to whether additional resources should be made available in order to fund the introduction of an intervention.

2.2.4 HEALTH INEQUALITIES

Eliminating health inequalities is considered as an important priority in public health. For example, the Netherlands has a research-based health inequalities tackling strategy.[7] This strategy is developed by an independent committee and is intended to reduce socioeconomic inequalities in disability-free life expectancy by 25% in 2020. Further, the United Kingdom set reduction of health inequalities as a key aim of health policy.[8] Evidence and expert judgments on areas suitable for policy development formed the basis of a plan of action to reduce health inequalities. Finally, the WHO Regional Office for Europe (WHO/Europe) published a review of social determinants of health and the health divide to address health inequalities within and between countries across the 53 member states of the European region through appropriate public health policies (http://www.instituteofhealthequity.org/projects/who-european-review).

2.2.5 MONITORING AND EVALUATION OF PUBLIC HEALTH POLICIES

Monitoring and evaluation is an essential step in the public health policy development. It provides information about the value for money of the policies and about possible improvements when the policy is revised. In the current climate of constrained financial and human resources, evaluations help in the development of effective policies. Some existing programs such as childhood vaccination or tobacco control have been evaluated, whereas other programs that are newly introduced are still not evaluated. If an intervention is unsuccessful, the evidence should help to determine whether the intervention was inherently faulty (failure of concept or theory) or badly delivered (failure of implementation).[9] The appraisal of evidence about public health interventions should encompass not only the credibility of evidence, but also its completeness and its transferability. This area requires more attention and research effort.[10]

2.3 DECISION-MAKING ON THE REIMBURSEMENT

Making decisions on the reimbursement of new drugs, interventions, and services is among the most important responsibilities of decision makers. Here, we focus on publicly funded health care systems, regardless of whether they are funded via taxation or social security contributions.

It is a complex process from which several issues can arise. The main issue relates to the fact that resources for financing health care are always finite. Therefore, expenditure must be rationalized, and priorities must be set. Priority-setting is undertaken at different levels, from resource allocation to the assessment of the value of the interventions. In those circumstances of scarcity, decision makers will be guided by the notions or fairness and justice.

More importantly, criteria for public health decisions need to reflect the values shared by the majority of people in a given society. This is a core concept, because these values may differ across societies (countries). For example, some societies believe that wealth spent on health care should be shared across all citizens according to their individual needs. This is regardless of their income, age, and life expectancy in their current health condition, productivity, or the rarity of their disease. This is called equity in access to health care.

In contrast, other societies believe that the distribution of wealth on health care should be allocated equally (or *fairly*) for everyone, regardless of their need. Consequently, people who need excessively costly health care cannot be reimbursed from public funds, because they cannot receive more wealth than an average citizen. This means that people who suffer from diseases for which only costly treatments exist will have to pay for them out of pocket or from other than public sources. This is called equality in access to health care.

Therefore, priority setting is related to values shared by a given society. Methods of eliciting these values are described in Section 2.4.

Daniels[11] points out: "when we knowingly and deliberately refrain from meeting some legitimate needs, we had better have justification for the distributive choices we make." Therefore, social value judgments are required to set the priorities.[12] These values include the following:

1. Accountability for reasonableness
2. Transparency
3. Participation

2.3.1 ACCOUNTABILITY FOR REASONABLENESS

Daniels and Sabin[13] developed an ethical–theoretical framework for accountability for reasonableness and the requirements for a legitimate decision process:

- Transparency
- Relevance of decision criteria
- Revisability of decisions (revision)
- Enforcement of the transparency
- Revisability and relevance conditions

Relevance of the criteria means that the criteria used for decision-making are supported by the society. This does not imply that the majority view will always be followed. Transparency in the application of the criteria means that the criteria weights are explicit and were determined by a clear process. Revisability means that when new evidence becomes available or preferences or values change, decision makers should revise reimbursement decisions.

Gibson et al.[14] (Figure 2.1) developed a model to help decision makers make their priority-setting. This model helps put the concept of accountability for reasonableness into practice; it integrates empirical realities on how decisions are made and ethical values on how decisions should be made. Rationales are the factors determining a decision (safety, effectiveness, and cost-effectiveness) and reasons for the decision. The

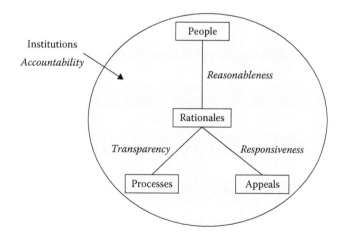

FIGURE 2.1 Model of accountability for priority setting. (From Gibson, J.L. et al., *BMC Health Serv Res*, 2, 14, 2002.)

contribution of experts and stakeholders in the rationales improves the reasonableness. The possibility to appeal can improve the responsiveness.

2.3.2 TRANSPARENCY

Priority-setting by definition means giving a higher importance to something over another. In public health, this means that some legitimate needs will be inevitably unmet; this is why it is controversial and transparency is highly recommended. Transparency consists of having reasons and criteria for the decisions that are explicit and relevant.[15] This can help patients and clinicians accept decisions as fair. If the decision-making processes allow decisions to be compared with previous decisions, the rationales will be more transparent. Transparency may depend on increasing accountability or participation.

Transparent decision-making process suggests that the possibilities of resources misuse, political biases, or clinical decisions are minimal, and the decisions are more likely fair and equitable. In order to insure the value of transparency, a revision of the decisions should be made openly to confirm that decisions are not a "result of power plays behind closed doors."[16]

2.3.3 PARTICIPATION

In democratic countries, citizens have equal right to participate in the way they are governed. Involving the public, patients, and health professionals in priority-setting may be a way to improve those decisions, promote trust and confidence in decision makers, and improve transparency. Potential users or users of health services can communicate the user point of view to decision makers that will assess the services; they can bring important input via their direct experience of drugs or services. On the other hand, it has been argued that a wider view should be adopted and the perspective of not only the patients but the citizens as well should be engaged.[17] However, the inclusiveness should be balanced, as including too many people can make decision-making cumbersome.[18]

Engaging public in decision-making can be made using several mechanisms: surveys, public consultations, community forums, citizens' juries, and deliberative polls. Mort and Harrison[19] defined two dimensions for those mechanisms: information and deliberation.

- Deliberation like a focus group is where participants discuss and interact about the questions.
- Consultation like public consultation is nondeliberative where participants respond to questions without discussion opportunities.

These approaches face some difficulties for application; for example, how to choose a representative sample of the population that can be statistically significant without being too large making a genuine deliberation impossible and how to avoid selection bias.

2.4 METHODS OF INCORPORATING SOCIETAL PREFERENCES INTO DECISION-MAKING

As mentioned before, decision-making should be guided by societal values. In collectively funded health systems such as those in England, cost-effectiveness has been the criterion used to guide reimbursement decisions. The English NHS is framed by a defined, allocated budget, so rationing of resources became necessary. Economic evaluation-based health technology assessment (HTA) was accepted as a way to rationalize rationing in a system where efficiency of the care provided is a major concern for their suppliers and recipients.[20]

In practice, using this criterion might face important real-life complications.[21] The basis of the estimation of quality adjusted life years (QALYs) relies on equal value given to all QALYs. However, society may give a higher value for QALYs gained by certain health problems or certain types of patients.[22] Furthermore, society may place value on aspects beyond the quality of life measured by QALYs such as improved productivity, happiness, and reductions in health inequalities. In recognition of this, HTA bodies started taking into consideration other criteria and society preferences alongside cost-effectiveness.

To elucidate the population preferences for a specific intervention, a multicriteria decision analysis (MCDA) is used. Devlin et al. defined MCDA as a set of different approaches, technical and nontechnical, to aid decision-making where decisions are based on several criteria. MCDA makes explicit the relative importance of the impact of every criterion on the decision and facilitates transparency and replicability in decision-making.[23] MCDA can help to structure the process and make it more consistent, but requires criteria weights and criteria scores to be elicited using, for example, discrete choice experiments (DCEs).

Illustratively, the national Belgium HTA body conducted a general population-based DCE that concluded that disease severity in terms of quality of life under current treatment and opportunities for improving quality of life through health care interventions are considered to be the most important criteria for resource allocation decisions in health care.[24] Compared to the decision makers, the general public attached relatively less importance to changes in life expectancy.

MCDA methods have been developed to create a structured process. The first two steps of the process are (1) problem structuring and capturing evidence that aim to identify alternatives and (2) criteria by meeting all the relevant stakeholders, doing literature reviews, defining the problems, goals, constraints, and uncertainties. The third step is the MCDA modeling; transparent mathematical approaches are used to measure the overall performance of the criteria. At the end, statistical analysis and sensitivity analysis are performed (Figure 2.2).[25]

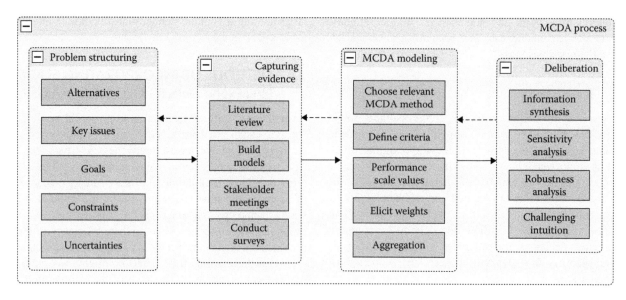

FIGURE 2.2   MCDA process. (From Thokala P. Multiple criteria decision analysis for health technology assessment, Report by the decision support unit, School of Health and Related Research, University of Sheffield, Sheffield, UK, February 2011.)

REFERENCES

1. Report of the Premier's Advisory Council on Health (Donald Mazankowski, Chair). A Framework for Reform, 2001.
2. Health Impact Assessment (HIA). The determinants of health, WHO. Available from http://www.who.int/hia/evidence/doh/en/ (accessed September 28, 2016).
3. Allin S, McKee M, Mossialos E, Holland W. *Making Decisions on Public Health: A Review of Eight Countries*, London: European Observatory on Health Systems and Policies, 2004.
4. Health Policy Consensus Group. Options for Health care Funding. Retrieved from The Institute for the Study of Civil Society, London, UK. Available from http://www.civitas.org.uk/pdf/hpcgSystems.pdf (accessed September 28, 2016).
5. WHO Report. Approaches to priority setting, Chapter 3, background paper. Available from http://www.who.int/medicines/areas/priority_medicines/Ch3_Approaches.pdf (accessed September 28, 2016).
6. Massetti M. et al. A comparison of HAS & NICE guidelines for the economic evaluation of health technologies in the context of their respective national health care systems and cultural environments. *J Market Access Health Pol.* 2015;(S.1)3. Available from http://www.jmahp.net/index.php/jmahp/article/view/24966 (accessed September 28, 2016).
7. Mackenbach JP, Stronks K. A strategy for tackling health inequalities in the Netherlands. *Br Med J.* 2002;325:1029–1032.
8. Marmot M. Social determinants of health inequalities. *Lancet.* 2005;365:1099–1104.
9. Rychetnik L, Frommer M, Hawe P, Shiell A. Criteria for evaluating evidence on public health interventions. *J Epidemiol Community Health* 2002;56(2):119–127.
10. Framework for Program Evaluation in Public Health, CDC. Available from http://www.cdc.gov/mmwr/preview/mmwrhtml/rr4811a1.htm (accessed September 28, 2016).
11. Daniels N. Rationing fairly: Programmatic considerations. *Bioethics.* 1993;7:224–233. doi:10.1111/j.1467-8519.1993.tb00288.
12. Clark S, Weale A. Social values in health priority setting: A conceptual framework. *J Health Organ Manag.* 2012;26(3):293–316.
13. Daniels N, Sabin J. Limits to health care: Fair procedures, democratic deliberation, and the legitimacy problem for insurers. *Philos Public Aff.* 1997;26(4):303–350.
14. Gibson JL, Martin DK, Singer PA. Priority setting for new technologies in medicine: A transdisciplinary study. *BMC Health Serv Res.* 2002;2(1):14.
15. Butler J. *The Ethics of Health Care Rationing: Principles and Practices*, London: Cassell, 1999.
16. Daniels N. Benchmarks of fairness for health care reform: A policy tool for developing countries. *Bull World Health Organ.* 2000;78(6):740–750.
17. Coote A. Direct public and patient involvement in rationing. In New B. (Ed.), *Rationing Talk and Action in Health Care.* London, UK: BMJ Publishing Group, 1997, pp. 158–164.
18. Martin D, Abelson J, Singer P. Participation in health care priority-setting through the eyes of participants. *J Health Serv Res and Pol.* 2002;7(4):222–229.
19. Mort M, Harrison S. Health care users, the public and the consultation industry. In Ling T. (Ed.), *Reforming Health Care by Consent: Involving Those Who Matter.* Oxford: Radcliffe Medical Press, 1999.
20. Oswald M. It's time for rational rationing. *Br J Gen Pract.* 2013;63(612):e508–e509.
21. Birch S, Gafni A. Cost effectiveness/utility analysis. Do current decision rules lead us to where we want to be? *J Health Econ.* 2002;11(3):279–296.
22. Shah K. Severity of illness and priority setting in health care: A review of the literature. *Health Pol.* 2009;93:77–84.

23. Devlin N, Sussex J. *Incorporating Multiple Criteria in HTA. Methods and Processes*. London: Office of Health Economics, 2011.

24. Cleemput I, Devriese S, Kohn L, Devos C, van Til J, Groothuis-Oudshoorn K, Vandekerckhove P, Van de Voorde C. Incorporating societal preferences in reimbursement decisions—Relative importance of decision criteria according to Belgian citizens. Health Services Research (HSR), Brussels: Belgian Health Care Knowledge Centre (KCE). 2014. KCE Reports 234. D/2014/10.273/9.

25. Thokala P. Multiple criteria decision analysis for health technology assessment, Report by the decision support unit, School of Health and Related Research, University of Sheffield, Sheffield, UK, February 2011.

SUGGESTED READING

Clark S, Weale A. Social values in health priority setting: A conceptual framework. *J Health Organ Manag*. 2012;26(3): 293–316.

3 Definitions and Concepts

3.1 ORIGIN OF THE MARKET ACCESS TERM

3.1.1 Market Access for Goods

The market access (MA) concept was first introduced by the World Trade Organization (WTO) to define the competing relation between the domestic and the imported products of a country. The WTO was established under the Marrakesh Agreements of 1994. The main intents behind this organization were to promote free market, transparency, reciprocity, and nondiscrimination in the field of international trade between participating countries. As health is a particular kind of good, as we discussed in Chapter 1, we present the context in which pharmaceutical MA should be seen, in Sections 3.1.2 through 3.1.6.

The WTO defines MA as a set of conditions, tariff, and nontariff measures, agreed by members for the entry of specific goods into their markets, that is to say the government policies regarding trade barriers in general, and specifically the issues of import substitution (to promote local production) and free competition. These trade barriers are used to encourage and protect domestic industries from foreign competition, allowing as a consequence, regulating consumption and reliance on imports. However, in a situation where each country individually sets the rules for access to its own market, it can eventually lock it down, and proscribe all foreign competing products. This obviously affects each country's exports in a negative way in the context of an international trade system. WTO members continually seek to improve MA liberalization.

There are two types of trade barriers established by countries, which are as follows.

3.1.1.1 Tariff Measures

Tariff measures are taxes on imports of commodities into a country or a region. Tariff commitments for goods are set out in each member's schedules of concessions on goods. The schedules represent commitments not to apply tariffs above the listed rates—these rates are *binding*.

For example, in the case of agricultural products, these concessions and commitments also relate to tariff rate quotas, limits on export subsidies, and other kinds of domestic support.

3.1.1.2 Nontariff Barriers/Measures

Nontariff trade barriers (NTBs) are any measure other than import duties (tariffs) used to restrict imports. They come in many different forms such as direct and indirect price influencers and are dealt with under specific WTO agreements.

The typical direct price influencers are export subsidies or drawbacks, manipulation of the currency exchange rates, inaccurate methods of imports evaluation, high customs charges, and lengthy procedures as well as minimum import prices.

Indirect price influencers have more to do with technical regulations and licensing. Some countries require unreasonable standards and inspection procedures in order to discourage or to restrict imports.

Although tariff barriers have steadily declined over the past few years, NTBs, such as technical regulations, along with safety or sanitary measures, have been increasing. Officially, governments worldwide introduce more and more regulatory requirements in order to address public health, safety, and environmental issues. Also, import prohibitions, requirement of a distribution network or effective means of marketing, or of homologation for a product in a given country are NTBs.

Technical regulations are among the most common NTBs, and they can be particularly effective due to the gap between European and foreign regulations. Homologation procedures are closely related to technical regulations and can be as demanding. For example, motor vehicles imported to China must undergo a number of approval/homologation procedures conducted by often uncoordinated local authorities. Moreover, under the provisions of the 2002 China Compulsory Certification System, each automotive product or component that has been approved in Europe must again be tested in Chinese laboratories, increasing the onus on the motor vehicle's importer. Similarly, in European countries, standards for fuel-efficient and low CO_2 emission vehicles have been implemented, to secure a market share for the local manufacturers who have committed, in return, to the policy of carbon-emission reduction.

Trade barriers are classified in Table 3.1.

3.1.2 Application to Health Care

The concept of MA can easily be transposed to pharmaceuticals. For example, certification or homologation in the case of motor vehicles is equivalent to the marketing authorization that is necessary in the pharmaceutical field in order to access a new market. Remaining trade barriers mentioned earlier also apply to the pharmaceutical market.

3.1.2.1 Tariff Barriers on Pharmaceuticals

As most developing countries import pharmaceutical products, they charge import tariffs, value-added tax (VAT), and other domestic taxes on these products to generate revenue and protect the local manufacturers from competition. These measures have little or no impact on patient's access to the medicine in a country that runs a national health insurance system. This is the case of most developed countries that apply low or no taxes on pharmaceuticals. However, they have a direct impact in other countries where there is no such health insurance system in place where the local pharmaceutical

TABLE 3.1
Trade Barriers Classification

Type of Barrier	Classification
Tariff measures	• Import policies reflected in tariffs and other import charges • Import policies reflected in quotas, import licensing, and customs practices • Standards, testing, labeling, and various types of certification/homologation • Direct procurement by government
Nontariff measures	• Subsidies for local exporters • Lack of copyright or patent protection • Restrictions on franchising, licensing, and technology transfer • Restrictions on foreign direct investment, and so on

manufacturing is limited, and the burden of these charges is borne directly by the patient or a private insurance.

Nigeria, Pakistan, India, and China all have significant local industries and are among the group of countries with the highest import duties. These tariffs and taxes are set to protect local manufacturers who rely on high import barriers for their survival.

The global trend has however been to reduce or eliminate tariffs and taxes on medicines in order to stimulate trade, competition, and the scaling down of prices.

3.1.2.2 Nontariff Barriers on Pharmaceuticals

Nontariff measures have dramatically increased in the pharmaceutical field.

The pharmaceutical industry is one of the most highly regulated, with complex measures from marketing authorization, efficacy and safety controls, quality standards, pricing, and reimbursement of pharmaceutical products, to import and distribution regulations, and so on. Most governments have instituted mandatory procedures to ensure the safety and efficacy of the medicines distributed in their market. As a consequence, the pharmaceuticals trade is often subject to strict regulations emanating from local drug approval agencies. Although such regulations are needed to ensure that the society has access to safe and effective medicine, they may also be designed to protect local health care industry.

Requirements for drug registration may involve specific local clinical studies that are not necessarily legitimate. An increasing number of countries require clinical trials of a new product to be conducted or repeated on their territory. For example, in China and Russia, early clinical tests as well as pivotal trials need to be repeated locally, which may take up to 5 years. Vietnam requires foreign companies to conduct on-site trials for products that have been marketed in their country of origin for less than 5 years, yet it does not require the same from local manufacturers.

In Russia, a pharmacokinetic study or a bioequivalence study is requested on Russian population performed in the Republic of Russia in order to obtain marketing authorization. This leads to development of clinical industry in Russia that is funded by the international pharmaceutical industry.

Furthermore, many countries have prescribed that pharmaceutical companies follow local good manufacturing practices (GMP). Without mutual recognition of foreign GMP, these regulatory demands become burdensome and time-consuming procedures that undermine effective MA for medicines. For example, in Turkey a specific regulation for GMP was set up. This regulation differs from international standards, is specific to Turkey, and is only followed by Turkish manufacturers. In order to obtain Turkish marketing authorization, pharmaceuticals need to comply with the Turkish GMP. Only if no product compliant with Turkish GMP is available, other products may be approved. This rule protects the national generic industry from foreign competition.

In Tunisia, the national generic companies are not requested to provide bioequivalence studies, although it is mandatory for foreign generics.

Some countries create double standards with different regulations for nationally owned companies and foreign-owned companies. This is the case for China and Saudi Arabia for example.

Other countries deliberately delay the approval of foreign medicines to minimize competition with national products and/or delay expenditure on new drugs by the national health insurance. South Africa's Medicines Control Council (MCC) requires that all new medicines obtain its own regulatory approval before they can be marketed in the country, even though they have already been approved by reputable foreign regulatory bodies such as the FDA. This leads to an additional time of 39 months on average for drugs registered in the United States, Europe, and Japan, before being approved in South Africa.

Finally, an interesting example illustrates how the various types of barriers can be played around with by different countries. Before 2005, India was not a member of the WTO and the patent protection was not observed in this country. Drugs could be manufactured and sold on the Indian market without the patent holder's consent. In turn, to protect their own industry, other WTO members set high custom fees on Indian generics that were already off patent outside India. Consequently, to sustain their own industry, India had no choice but to enter the WTO and respect the patent rules in order to benefit from the open market conditions for their pharmaceutical products.

Finally, achieving positive reimbursement recommendation in countries with significant health insurance has become the most complex obstacle pharmaceutical companies need to face.

3.1.2.3 Health Care Market Specificities

In spite of many similarities between health care products and other goods in a free market economy, the former is a unique field that challenges the traditional economic paradigm. First, health is a nonmarket good in that it cannot be traded or borrowed. Medicines, however, are market goods intended to improve the health of their consumers or, in other terms, to produce health. Therefore, they can be sought and consumed as a proxy for health but not as health.

However, unlike other goods that obey the supply and demand model, prescription medicines' price is often determined by the institutionalized payers through negotiation with the manufacturers or is simply set by the manufacturer. Therefore, medicine prices do not reflect accurately the demand, supply, or the costs of production. Another characteristic is the inherent uncertainty to the consumption of medicines. In a traditional market context, the utility of a product is evident, and the benefits can be obtained often immediately after consumption. However, when it comes to medicines, the health benefits cannot be achieved immediately and are often uncertain. As it happens, the effects of drugs differ from one patient to another. The health care market is also peculiar because of the unique relationships between the different actors of this market. In a traditional market economy, the consumer is also the buyer and the payer. They have a unique objective of increasing their utility when they buy goods. But in the health care market, payer, buyer, and consumer are three different stakeholders with different perspectives and objectives. The payer is the public or private health insurance system or the government, for whom the goal is to obtain the health benefits of a medicine for the patients although containing health care cost within a predefined budget. The buyer is the physician who prescribes the medicine, and whose interest is both to improve the patient's health and to maximize their revenue when providing care. Finally, the consumer is the patient who aims to maximize their own health but has no authority to access prescription medicines.

> If one is about to purchase a car, he or she will choose it, pay for it, and then drive it. Therefore, a single person will make a trade-off between the car's price and its attributes. On the contrary, in the case of medicines, the doctor decides which drug to prescribe, the patient will use it, and the health insurer will pay for it.

Further, there is an important asymmetry of information between the consumer and the buyer. The consumer lacks knowledge about pharmaceuticals and has usually no choice but to trust the physician. This knowledge gap decreased dramatically with the introduction of the Internet and science communication, as patients are increasingly better informed about various health conditions, medical procedures, and treatments at their disposal.

Another aspect of the health care market is high regulation. Marketing authorization for each new drug is a well-established requirement. Further, to protect the patients, prevent possible abuse and overprescriptions, physicians are increasingly enjoined to follow guidelines written by commissioned experts and approved by health authorities, so they are not fully free to prescribe any drugs they want. Also, indications for each new drug are increasingly precisely defined in its market authorization, and doctors may not always have the freedom to prescribe drugs according to their own experience and judgment.

Even if patients cannot always choose the medicine, they could in principle be able to change their doctor, in order to obtain a treatment that suits them the most. However, the rules of free competition are rarely observed in the health care market. Depending on the country and the health care system, patients may not always be free to choose a physician or a hospital. Also, depending on the partnerships arranged by the health insurance provider, they may not be able to switch their care provider, even when they are clearly unhappy with the service provided.

When it comes to decisions on drug reimbursement by the institutionalized payers, the consumers are rarely involved in the process. They cannot decide which treatment will be reimbursed, even if they are directly concerned by the medicine in question. However, there seems to be a trend of fostering an indirect participation of patients in that process. For example, in the United Kingdom, a representative of a relevant patients association is invited to take part in the National Institute for Health and Care Excellence's (NICE) Technology Appraisal committee meetings. This example is not specific to NICE but exists in most HTA organizations.

Finally, certain medicinal products can be acquired for virtually all potential consumers in a given country through direct procurement by the government.

In conclusion, there are four features that clearly differentiate the health care market from other markets:

- The price is not determined by supply and demand
  In a traditional market economy context, the price is determined by supply and demand. A single entity assumes the functions of the buyer, the payer, and the consumer. In the health care market, however, the prices are determined by payers through negotiation or are simply notified by the manufacturer. The buyer is the physician who prescribes the treatment, the payer is the health insurance provider, and the consumer is the patient. The three parties do not necessarily have convergent views on how health care should be delivered.
- Payers are committed to purchase health for the society
  The payer's intent is to provide health for the patient. When payers fund medicine they fund health production. They can only buy proxy of health through the purchase of medicine and health care services. The actual outcome in terms of health improvement remains uncertain.
- Health is specific to each individual
  Unlike food, real estate, or technology, health cannot be shared or traded between individuals. The outcome of a treatment or a procedure also depends

on individual characteristics of the patient. The patients' characteristics may not be fully known a priori because of the lack of appropriate tools. This repertoire of scientific tools is evolving fast and changes regularly our understanding and approach to disease and therapies.

- Externality of health

  Medicines can have a positive impact on the health of people, other than those who consume it. This is particularly the case for vaccinations and antibiotics. The treatment and prevention of contagious diseases at the level of an individual can protect the global population from a potential epidemic.

  - Restricting access to health care for a subgroup of the population can have dramatic impact on that population's health status.
  - Poor health care in a subgroup of the population will affect the health of the remaining part of the population that has good access to health care.

  This is one of the main reasons for the creation of national health care systems.

  Illustratively, it has been iteratively reported that despite the highest per capita health care expenditure, the United States does not have the best population health status, because of the wide disparity in access to health care.

Some of the seminal features of the health care market recently have undergone important changes (Table 3.2). The impact of these changes remains to be seen in the years to come.

3.2 MARKET ACCESS KEY CONCEPTS

3.2.1 WHAT IS VALUE?

MA is related to the concept of value for money, from a payer's point of view. As a result, the primary objective of MA studies is to define and measure the value of health services.

Value is an intricate concept present in a variety of disciplines:

- Philosophy
- Mathematics
- Sociology
- Accounting
- Economics

Value has different meanings in each of these fields of study. In philosophy, value can be defined as the importance of a moral or aesthetic judgment regarding a personal or social standard of conduct, morality, ethics, politics, spirituality, or aesthetics. In mathematics, value can generally be defined as the determination of a variable with notable variations: an approximate value of a number is a number for which the difference is small enough to be omitted in numerical applications and the absolute value of a number is the numeric value, regardless of its sign. In sociology, the values are the moral principles of a constitutional philosophy, which is ranked depending on the characteristics of the individual or the society.

The large variety and disparity of underlying concepts behind the word value is a source of confusion when speaking about value.

The term *value* is widely used in economics and accounting, as exemplified in Table 3.3.

In economics, value is a concept that refers to two different theories (Menger 2007). The first one is an objective theory, or the intrinsic theory of value, where the value of an object, good or service, corresponds to the cost of the production, that is the cost of raw material and human work needed.

The other one is subjective and is more consistent with the idea of value as perceived in the health care market. This theory of value advocates the idea according to which the value

TABLE 3.2
Recent Trends in Health Care Delivery Organization That Can Shift the Market Access Paradigm

Organization of Health Care Delivery

Until Recently	Emerging Trends
Asymmetry of information between prescriber and patient on the disease and on possible interventions	Patients are increasingly informed about their condition and possible treatments, mostly because of the widespread use of the Internet
Low or no involvement of patients in reimbursement decision-making process	Patient associations became more politically active and with lobby for public funding of (often costly) treatments that can improve their condition
Freedom to choose a doctor/hospital where the patient wants to be treated within the public health care system	Increasingly patients need to prioritize one GP and/or primary care unit and hospital, otherwise they might receive smaller or no reimbursement from the insurer
Prescription guidelines issued by specialist doctors' associations—doctors are not obliged to follow	Prescription guidelines issued by a national HTA body—doctors are obliged to follow and might have less freedom to choose treatments

TABLE 3.3
Various Applications of the Term "Value" in Economics and Accounting

Acquisition value	Customs value	Net value
Shareholder value	Exchange value	Residual value
Present value	Out of tax value	Value at risk
Net present value	Intrinsic value	Speculative value
Added value	Liquidation value	All taxes included value
Traded value	Market value	Marxist theory value
Gross value	Security value	Use value
Book value	Sales value	Work value

of a good is neither determined by any inherent property of the good, nor by the amount of labor required to produce the good, but is determined by the importance an acting individual places on a good for the achievement of their desired outcome. The price offered is not a measure of subjective value; it is just a means of communication between the buyer and the seller.

As far as health care and MA are concerned, this last definition is the most relevant and should be used. In MA, the value of a drug or a health service depends on the subjective perception of the payer regarding the medical need in the society and how the product addresses that need.

This assessment of value made by payers is rather subjective, yet based on scientific evidence, such as clinical trials, epidemiology, or cost-effectiveness studies. Most institutionalized payers formally require drug manufacturers and health care providers to submit evidence that corroborates the value of their product in terms of clinical outcomes and/or the cost of achieving such outcomes. Achieving a positive coverage decision, and so MA for a product, depends on the ability of the pharmaceutical industry to submit pertinent evidence. This calls for a thorough understanding of this evidence-based concept of value on the part of this industry.

The kind of evidence required by the payers for the assessment of a product differs from one country to another and covers a wide array of indicators, such as proof of clinical and economic value and more specific considerations of ethics, equity, and/or politics. The focus of the payer is always on assessing the value for money of a product.

The set of evidence generated and presented by the manufacturer for the payer will form the value proposition. This term of value proposition is often used in health care economics. The development of such proposition is the ultimate aim of MA activities from an industrial perspective.

However, from a payer's perspective, the objective is to relate the drug's value to the right price considering all evidence. This is one of the most debated issues at the moment among various health care actors, and is often called value-based pricing.

Value-based pricing is sometimes confused with cost-effectiveness-based decisions. Although the consideration of efficiency through cost-effectiveness model can be a part of the value assessment by payers and can help them set the right price, the value-based pricing concept is larger. It extends to clinical, ethical, and political considerations. Market access agreements (MAA) and health economics models are some of the various means of approaching this issue. The value-based pricing will be specifically addressed in Section 3.2.3.

3.2.2 WHAT IS ACCESS?

It is crucial to differentiate access from accessibility and MA. These are three different concepts that are often misunderstood.

- Access to health care or to health services is the perceptions and experiences of people as to their ease in reaching health services or health facilities in terms of location, time, and ease of approach. Lack of access: when a medicine is unavailable, inaccessible, or unaffordable.
- Accessibility is the aspect of the structure of health services or health facilities that enhance the ability of people to reach a health care practitioner, in terms of location, time, and ease of approach.
- Market access is the ability for a drug to achieve through a health insurance system, a reimbursed price, and a favorable recommendation for prescription.

Further, the stakeholders of the health care market have different objectives regarding access to medicine. Although the objective of the industry is to provide the largest possible access to their drugs, the objective of the payer is to restrict access to the most beneficial patients alone, in order to achieve the highest effectiveness and cost efficiency. The industry must persuade the payer of the medical relevance and value of the drug to obtain access to a larger target population.

3.3 MARKET ACCESS DEFINITION

If we consider the MA definition by the WTO, obtaining MA should be the ability to access the whole market in a given country, sell, and achieve revenue on this market without significant obstacles.

In cases of pharmaceuticals, the obstacles are the marketing authorization; the price and reimbursement (P&R) levels, logistics (storage and supply conditions), the drug surveillance (follow-up on potential and actual product adverse effects), and so on. In practice, however, the pharmaceutical industry has become proficient in addressing all those hurdles except P&R. Thus, MA has become synonymous to the hurdle of achieving high price and reimbursement levels.

However, pharmaceutical markets can have a varying degree of fragmentation, from countries with single national insurer to countries with multiple private insurers, or a mix of both. In the latter two cases, MA is the ability to systematically gain access at optimal conditions in each and every catchment area with each and every insurer. Depending on the type of health care market organization (e.g., centralized vs. decentralized or fully fragmented), the concept of MA may focus on different aspects.

Therefore, the concept of MA is heterogeneous and hard to define, depending on whether we are dealing with a private, public, or mixed health care system. In general, it is the process by which a company gets a drug available on the market after obtaining a marketing authorization and by which the medicine becomes available or affordable for all patients for whom it is indicated.

The following definition will be used in this book:

MA for pharmaceuticals defines the ability for a drug to achieve through a health insurance system a reimbursed price and a favorable recommendation for prescriptions.

It covers a group of activities intended to provide access to the appropriate medicine for the appropriate group of patients and at the appropriate price.

Also, we can define MA by describing the following:

- The objectives
- The actions
- The field activities
- The scope
- The process

The ideal outcome of MA is to achieve the optimal price with maximum reimbursement for the approved target population with no limitation on prescription or funding procedures. However, in practice there is a trade-off between the following:

- Price and reimbursement conditions
- Target population selection
- Prescription and funding procedures

Therefore, MA can also seen as activities that support the management of potential barriers, such as nonoptimal price and reimbursement level, the restriction of the scope of prescription for a drug, complicated prescription, or funding procedures.

Example: A drug belonging to the selective serotonin reuptake inhibitors class, which obtained a marketing authorization in several indications such as depression, social anxiety, panic disorders, obsessive–compulsive disorder (OCD), eating disorders, and chronic pain may be granted by payers a full reimbursement status only in two of these indications: depression and panic disorder. Consequently, the drug will have limited market access because it is not reimbursed for all patients in which it could be effective.

Moreover, market access can be further limited by payers if prescription of the drug is restricted to a narrowed population of patients (e.g., as second line of treatment in nonresponders to a first line drug or only for patients with a severe condition). Limiting prescribers to specialist doctors or putting in place a complex reimbursement scheme that imposes administrative burden on doctors or hospital pharmacists can hinder market access for a drug even more.

The scope of these activities is overlapping with the management of pricing and reimbursement, health technology assessment (HTA), and formularies. The formularies are the lists of medicines that may be prescribed at the expense of the institutionalized payer.

The MA process should be well understood by the different actors on the health care market. In Europe, the process can be outlined as follows (Figure 3.1):

Marketing authorization from a regulatory agency, which could be the FDA in the United States or European Medicines Agency (EMA) in Europe, is issued based on consideration of the product's safety, efficacy, and quality in the highly controlled conditions of randomized controlled trials (RCT). In the case of European Union (EU), national agencies are responsible for the implementation of this authorization in their local settings. Once a medicine is approved for marketing, HTA bodies are responsible for assessing its real-life efficacy (i.e., effectiveness), cost-effectiveness, relative efficacy, related medical need, budget impact, and other evidence that will be later used by payers for P&R decisions, as well as formulary listing and prescription guidelines.

Notably, EUnetHTA has developed a 9-domain Core Model® for HTA that includes the health problem and current use of technology, description and technical characteristics of technology, safety, clinical effectiveness, costs and economic evaluation, ethical analysis, organizational, social, and legal aspects.

Payers themselves are not qualified to evaluate those criteria, so they delegate these activities to independent groups of experts that produce the HTA evidence. HTA evaluations aim

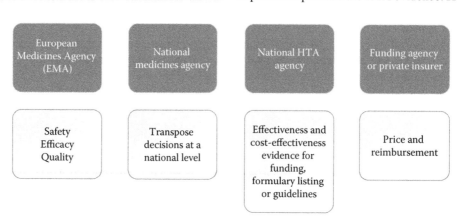

FIGURE 3.1 The various actors of the European health care market and their roles. Similar setups exists in other geographies.

to inform payers' decisions and help them set the appropriate P&R conditions.

As explained above, MA should not be confused with the following activities:

> What MA is not:
>
> - MA *is not* about obtaining regulatory approval (license, Marketing Authorization)
> - MA *is not* about medical representatives getting access to doctors or pharmacists (sales force)
> - Access to pharmacy shelves (distribution) *is not* MA
> - MA *is not* about choosing the right channel to promote product (e.g., marketing, direct-to-customer [DTC] advertising)

In more detail, MA requires a set of skills, different from those needed in development and regulatory affairs (Table 3.4).

These skills are the ability to *negotiate* with the institutionalized payers, rather than fulfill predefined criteria. Consequently, achieving full reimbursement often means being able to *make trade-offs* between a high price and an optimal one. Also, it involves convincing payers that *the level of uncertainty* around the drug effectiveness and/or cost-effectiveness in real-life use is acceptably low. Finally, in contrast to the centralized regulatory affairs, MA is a national to local activity, and thus requires adaptive tactics. Finally, where MA deals with payers, regulatory affairs deal with regulators. As shown in Table 3.5, regulators and payers have very different perspectives and requirements.

Likewise, similarities between MA and marketing activities can be misleading (Table 3.6). MA is an *evidence-based* discipline that requires generating and communicating scientific evidence. Institutionalized payers are *highly price-sensitive* and are held *accountable* for their decisions. Furthermore, they act under the influence of *multiple stakeholders,* and any successful MA strategy must recognize these influences. Finally, HTA authorities will require submission of *evidence of relevant superiority for positive recommendation,*

TABLE 3.4

Different Sets of Skills Required in Regulatory Affairs versus Market Access Activities

Regulatory Affairs	Market Access
Fulfill the requirements of market authorization	Negotiate with payers
Meet criteria for efficacy, safety, and quality	Determine trade-offs between price and MA to achieve optimal return on investment
Deal with certainty	Deal with uncertainty
Transparent regulation	Not transparent, fast changing rules
Global	National to local

TABLE 3.5

Different Perspectives and Requirements of Medicine Regulators and Institutionalized Payers

	Regulators	Payers
Internal versus external validity of data	Strong emphasis on internal efficacy focus (i.e., *well defined patient population*)	Emphasis on internal validity but with an external validity focus
Type of data preferred	Mostly RCT information	RCT, meta-analysis, observational data, registry, databases, and modeling
Endpoints preferred	Patient-relevant endpoints, clinical scales, and biological measures	Quality of life endpoints, number needed to treat, number needed to harm, functional measures, and so on
Scope of evaluation	Benefit-Risk ratio of specific product under evaluation, compared to placebo	Relative efficacy or effectiveness (in the disease area), compared to existing alternative treatments

TABLE 3.6

Differences between Market Access and Marketing Activities

Marketing	Market Access
Perception based	Evidence based
Audience not accountable	Price sensitive audiences accountable
Opinion leaders are key	Influenced by multiple stakeholders
Innocent until proved guilty	Guilty until proven innocent

otherwise they will not recommend payers to fund an intervention at a premium price.

3.3.1 Market Access and the Structure of Health Care System

Further, MA strategies depend on whether we are dealing with a private, a public, or a mixed health care system.

3.3.1.1 Publicly Funded Health Care Systems

Within publicly funded national health insurance found in most countries in Europe, the government defines the overall public health goals and corresponding funding usually through the parliament health budget vote. Then, the rules for access to the market for the industry are laid out by a central agency or agencies. These involve the evidence requested for the assessment of a product and the criteria for making the funding decision. The public health care payers represent the society interest and try to integrate the society perspective when making decisions.

3.3.1.2 Mixed or Private Health Care Systems

There are countries where the health insurance is fragmented and largely private, as in the case of the United States. There is no unified framework for obtaining MA in the United States, and the public and each of the private insurers follow their own pathway. In this setting, private health care payers engage in independent negotiations with the industry. This can be seen as a negotiation between two business entities that are looking to maximize their profits. However, in the United States, the public payers (the Centers for Medicare and Medicaid Services [CMS], e.g., Medicare, Medicaid, and the Children's Health Insurance Program [CHIP]) represent an increasing proportion of the health care budget that is about to match the commercial one. The CMS pathway resembles that of many European countries, Australia, or Canada, except that formal health-economic analysis or HTA is not compulsory in the United States, except in very rare cases. Further, high cost should not be a cause for a negative reimbursement advice by the CMS.

3.3.1.3 Centralized and Regional Market Access

A trend toward decentralization is emerging also in the public health care settings, as policy-making is increasingly devolved from the national bodies to local health authorities. As health care payers are compelled to curb their pharmaceutical budgets, in a context of economic stagnation, local policy makers are also exacted to decide on which therapies are funded, and under what conditions. However, these responsibilities are not always matched with competences at the regional level. In many countries, the regional authorities accountable for medicine spending are seldom prepared to negotiate the costs of the drugs or to assess their value. Concomitantly, there is increasing incitement to concentrate on cost-containment of the health care budgets they hold.

This trend is blurring the traditional division between countries with decentralized health care systems, such as Spain, Italy, Sweden, or Germany and countries with more centralized ones, like France or England. For example, in England, where strategic decisions affecting the national health system remain in the authority of the national department of health, the power of execution is assigned to a large number of primary care trusts (PCTs). Each PCT is responsible for the provision and funding of health care services for populations that range from 90,000 to over 1,259,000 individuals. Concretely, this means that, apart from the national bodies, the pharmaceutical industry has to engage directly with PCTs, in order to access the markets in England.

Infamously, this structure of health care delivery in England led to geographic inequalities in access to certain expensive life-saving therapies. They were included in the local health care baskets of some PCTs, but excluded from others, a phenomenon called *postcode lottery* by its critics. Although the role of NICE is to define the national health care basket, there is still scope for PCTs to include innovative therapies beyond the national requirements

defined by NICE. Even in France, a country with a very strong political history of centralization, recent reforms have devolved significant budgets and decision-making powers to 22 regional health agencies (*Agences Régionales de Santé* [ARS]).

As their regulatory responsibilities extend, regional health authorities increasingly deploy a variety of management and policy tools at their disposal. Such policies can be either directed at the industry, the medical community, or patients and can have the form of local market endorsement, prescription guidelines, drug formularies, prescribing incentives, or volume restrictions, patient deductibles, regulation of relations between doctors and industry representatives. To some extent, this creates a *double hurdle* for companies, as a favorable national decision needs to be replicated from the central to a regional level, with payers who have different concerns, budgets, and levels of sophistication as compared to the national ones.

The increased power of the regional bodies commands that the industry must be fully aware of the local requirements along the national ones. Further, in a decentralized country, there can be great disparities between regional health care budgets, general wealth, public health priorities, demographic, and epidemiological characteristics. This calls for strategic attention from pharmaceutical companies. Ideally, they should adopt tailor-made MA strategy for each region. However, costs of local adaptation can be high and local requirements for data analysis and support can be very specific.

For example, different regional payers may use different methods to evaluate health technologies. Some rely solely on the budget impact of new products, whereas others on the cost-effectiveness based on local epidemiology, costs, and clinical practice. E.g. in Italy, 16 of the 21 regions use national drug formularies, whereas eight regions undertake some sort of local HTA. Consequently, the access to regional markets cannot be achieved with a *one size fits all* approach and necessitates on the part of the companies the understanding of the particular features of the regions they wish to access.

3.4 CULTURAL SPECIFICITIES OF MARKET ACCESS

Finally, the MA strategy needs to be culturally-sensitive, even among countries that seemingly apply the same methodology to inform their drug funding decisions. For example, countries that employ formal HTA can still substantially differ in the objective, the process, and the impact of the HTA in MA (Table 3.7).

Further, the structure of the MA environment differs among countries and needs to be mapped in order to prepare for a successful strategy (Table 3.8).

Finally, there is a geographical dichotomy between the northern and southern European countries, the former being more centralized and reluctant to price negotiation (Figure 3.2).

TABLE 3.7
Cultural Differences between Countries Regarding the Objective, the Process, and the Impact of HTA Evaluation in Market Access

	France	Germany	United Kingdom
Objective	Secure access to all new products, but at the right price	Obtain savings on drug spending with no detriment to safety/efficacy	Obtain rational allocation of resources
Process	*Driver*: Public health relevance of benefit compared to the next best alternative	*Driver*: Same effect—same price (e.g., jumbo groups)	*Driver*: Maximization of efficiency of the health care output
	Method: Single double blind randomized clinical trial effect size	*Method*: Meta-analysis efficiency frontier as a backup	*Method*: Cost utility threshold is £30,000/QALY
Impact	Gate keeper for market entry	Reimbursement level	Recommendation for prescriber formulary listing

TABLE 3.8
The Various Elements of the Market Access Environment Present in Different Countries

	United Kingdom	Germany	France	Italy	Sweden	Spain	Canada	Australia
Pricing policy			✓[a]	✓[a]	✓[a]	✓	✓	✓[a]
Reimbursement board			✓[a]	✓[a]	✓[a]		✓	✓[a]
National HTA	✓	✓	✓		✓		✓	✓
Regional formulary		✓		✓	✓	✓	✓	
Local formulary	✓	✓				✓	✓	

[a] Linkage between pricing and reimbursement

Northern Europe (UK, Scandinavia, The Netherlands)
Prescribers follow guidelines and recommendations form the National Health Insurance/Authority and are willing to accept cost containment measures
Payers use cost-effectivness to support decisions and may use restriction of drug recommendation to subpopulations of patients

Southern Europe (France, Italy, Spain)
Prescribers behavior is difficult to control by the National Health Insurance/Authority
Payers use efficacy to support decisions and prefer to negotiate lower price with manufacturer rather than restrict drug use to subpopulations of patients

FIGURE 3.2 The cultural differences among prescribers and payers in Europe.

3.5 MARKET ACCESS FOR PAYERS

3.5.1 PAYERS EMPLOY MARKET ACCESS TOOLS TO CONTROL DRUG EXPENDITURE

The continuous growth of health care expenditure and more specifically pharmaceutical expenditure has put health care insurance providers under escalating pressure. For payers, MA tools are a powerful way to control drug expenditures. Quotations given below exemplify the growing concerns about the cost of novel treatments in the past decade:

The near-doubling of the median survival (in colorectal cancer) achieved over the past decade has been accompanied by a staggering 340-fold increase in drug costs—just for the initial eight weeks of treatment.

Dr. Deborah Schrag
Memorial Sloan-Kettering Cancer Center, 22 July 2004

(Referring implicitly to bevacizumab) There is a shocking disparity between value and price, and it's not sustainable. The industry will bring about government price controls which will be devastating for the industry…The industry has a black eye, and the market will correct that.

Dr. Roy Vagelos
Former CEO of Merck & Co., Annual Meeting of the International Society for Medical Publication Professionals, 30 April 2008

Market structure effectively provides no mechanism for price control in oncology other than companies' goodwill and tolerance for adverse publicity...Oncology drug pricing is a key long-term overhang.

Dr. Steven Harr
Morgan Stanley Research Notes—June 2005
and 3 October 2006

Despite an increasing proportion of products for which cheaper generic versions exist, the pharmaceutical market value continues to grow. To tackle this growth, payers have employed a variety of cost containment measures since the late 1990s. Nevertheless, they failed to control the expenditure. In the Organisation for Economic Co-operation and Development (OECD) countries, excluding the United States, health care spending has almost doubled its share of gross domestic product (GDP) over the past 10 years. The demographic changes and the expected future innovations are expected to generate a disruptive pressure unless appropriate action is taken. Pharmaceutical growth is a lot more significant than the health care growth, and accounts for as high as 20% in many developed countries.

The most common regulation of drug expenditure is price control where the institutionalized payer decides on the appropriate price of a medicine after negotiation with the marketing authorization holder. Only two developed countries seem to still enjoy a free (uncontrolled) pricing process: the United States and the United Kingdom. However, the latter one has put in place a regulatory process that indirectly regulates prices, that is, if a drug is thought to be overpriced, the access to the market is narrowed by means of a negative or restricted financing recommendation. Free pricing in the United Kingdom was supposed to be replaced by a controlled pricing process, following the recommendation of the United Kingdom's Office of Fair Trading. Although the initiative of value-based pricing failed, the new way to control product prices became to accept very high discounts that remain confidential but are often above 50% of the listed price.

Other pharmaceutical cost-containment measures developed by payers include general price cuts or exceptional taxes on turnover and profit.

The objectives of these payer activities are to

- define and allocate an acceptable health care budget according to health objectives and available resources.
- implement transparent rules to set fair price for medicines.
- ensure that highly priced medicines are used only within the authorized medical condition.

During the 1990s, the pricing regulation in Europe was often based on the health authorities' subjective perception of what the right price was. In order to dissolve political pressure around patients' access to new medicines and incentives for the industry to innovate, the authorities needed to implement more clear and objective rules for establishing prices. This resulted in the following two key developments:

- The creation of reference pricing within therapeutic class and across European countries
- The creation of national HTA bodies across European countries, Australia, and Canada that assess evidence supporting the benefit of new medicines and other health technologies

This trend is also seen in the United States where the American Recovery and Reinvestment Act (ARRA) provided $1.1 billion for comparative effectiveness research. The Federal Coordinating Council for Comparative Effectiveness Research (FCCCER) aims to support health care policy decision makers by generating research that involves large-scale pragmatic trials, patient databases, and the development of new quantitative methodologies. Unlike in many European countries however, outcome of this research could not be used explicitly by payers to deny access to a new technology. However, pharmacy and therapeutic advisory committees that review new drugs and make recommendation can use this research to guide their decisions. Also, Pharmacy dossiers of the Academy of Managed Care are essentially HTA recommendations. Notably, manufacturers are not obliged by law to disclose all information for the preparation of the dossier, and so the data may be skewed or selective to favor cost-effectiveness of a drug.

HTA is a process of evaluating the consequences of a new health care technology, as compared with products that are already available on the market. It summarizes information about the medical, social, economic, and ethical issues related to the use of a health technology in a systematic, transparent, unbiased, and robust manner. In order to generate this often sophisticated evidence, payers delegate the assessment to experts. As mentioned above, governments in most developed countries created HTA agencies that have the required expertise and that can act as independent stakeholders, not influenced by economic or political considerations.

National HTA process and corresponding drug funding decisions can have dramatic impact on the accessibility of the drug in a given country, as illustrated in the example below.

Example: Although national GDP is a good predictor of health care expenditure and patient's access to novel drugs, market access environment can modulate this dependency. For example, expensive drugs that are indicated in multiple sclerosis (MS) such as interferons and glatiramer acetate are fully reimbursed in most western European countries. However, in the United Kingdom, these drugs were not recommended by the United Kingdom HTA agency (NICE) in 2000 and were finally available to patients from 2002 within a complex risk-sharing scheme, where their prescription was restricted to specialist doctors. Consequently, only 12% of MS patients in the United Kingdom had access to the treatment.

However, despite these regulations and actions, the institutionalized payers continue to struggle to control drug expenditure. Indeed, when explaining the consequences of the growing use of evidence based medicine, United Kingdom's authors affiliated to the NHS said that "Some fear that evidence based medicine will be hijacked by purchasers and managers to cut the costs of health care. This would not only be a misuse of evidence-based medicine, but suggests a fundamental misunderstanding of its financial consequences. Doctors practicing evidence based medicine will identify and apply the most efficacious interventions to maximize the quality and quantity of life for individual patients; this may raise rather than lower the cost of their care" (Sackett et al., 1996).

3.5.2   How to Identify Payers?

In health care, payers are generally entities that finance or reimburse the cost of health products and services. We can also say that any *price-sensitive audience* is a payer. As an exception to this definition, social activists involved in medicine price campaigns could be considered price-sensitive audience, but not as payers because they do not finance or reimburse the cost of health services. In the health care market, payers always act as gatekeepers for MA.

Further, payers can have the following four pairs of characteristics:

- Directly or indirectly incentivized
- Decision maker or not
- Prescriber or not
- Acting for their own organization or not

In most European countries, there is one main payer in each country, corresponding to the national public health insurance or fund. Sometimes, there are additional payers at a regional level. There can be a mix of national and fragmented private payers as in the United States. More importantly, each payer can have different objectives, perspectives, and processes.

Depending on the country and level of authority, payers can be the following entities

- Members of national pricing committees (e.g., France, Italy, and Spain) and other key staff of the national health insurer
- Members of HTA committees: either national (the United Kingdom, Germany, etc.) or regional HTA bodies (Spain, Sweden, etc.)
- General practitioners in the United Kingdom and Germany, where doctors are rewarded for prescribing performance—their remuneration can be linked to cost-containing prescribing behavior
- Private health insurers (analogous to national insurers, but under somewhat smaller political pressure)
- Pharmacists (particularly chief pharmacists in the United Kingdom)

- Hospital managers and hospital staff with whom payers interact
- Employers who pay for health insurance plans

Payers should not be considered as a homogeneous audience but rather as a complex and heterogeneous one. The arguments accepted by one payer may be counterproductive for another payer within the same country. So the messages and the arguments should be tailor-made or at least customized for the targeted payer.

3.5.3   How Payers Assess Value?

Payers are concerned about the value of the medicine in order to contain drug expenditure and invest in products that can create best health outcomes. In this endeavor, they need to assess the uncertainty about the drug's potential health benefits, as well as the potential costs related to funding it.

The process of assessing the value for money of a medicine is broadly a four-step assessment, although not all payers go through the following four steps:

1. Comparative efficacy from clinical trials (as compared to alternative treatments for the same condition)
   It aims at comparing two drugs in clinical trials and measuring the benefit of one over the other. The clinical trial design, the inclusion/exclusion criteria, the randomization procedure, and so on may compromise the quality and reliability of the comparison and raise doubts for the payer on the actual effect size of the benefit
2. Comparative effectiveness from real-life data on use of the medicine
   If added benefit is observed in clinical trials, there are three potential obstacles for acknowledging the benefit in real life (effectiveness). Indeed, payers will have to address three uncertainties to be able to judge the product's value.
   - The effect size of the additional benefit in clinical trials: if the outcome of the trials is likely to be affected by bias, it remains uncertain if the actual effect size of the additional benefit in the trials will be attainable de-facto in real life.
   - The transferability across jurisdictions remains an important aspect to address, as different health care systems may treat and manage patients differently, making it difficult to extrapolate benefit from one jurisdiction to another.
   - Finally, the transferability from a clinical trial model to real life is also a major challenge. In clinical trials patients are highly selected and patient management is protocol driven, which does not match the real life practice.
     The appreciation of the effectiveness requires that these three specific uncertainties are addressed at least qualitatively and ideally quantitatively.
     Some countries stop their assessment at this phase.

If a medicine does not show significant benefit after these two steps, the value will be considered equal or lower than that of the comparator treatment. In this situation, no premium price can be granted. However, if the benefit is shown, value for money can be further assessed by comparing the extra benefits to the extra costs of the new medicine.

3. Cost-effectiveness

This methodology compares the effectiveness benefit against the cost consequences (e.g. cost per life year saved, per quality adjusted life year [QALY], per treatment success, per relapse avoided, etc.). Cost per QALY seems to have been increasingly adopted over the recent years in most HTA organizations. As resources are limited whenever a new intervention is introduced, the new one will displace the other available intervention (opportunity cost). It is important to consider if this opportunity is going to be at least as cost-effective as the one it displaced. Although it remains quite theoretical, as the effectiveness of intervention is often unknown at market launch, it is commonly considered rational to set a threshold for the incremental cost-effectiveness ratio (ICER) per QALY that represents the reference for available interventions that may be displaced.

4. Budget impact

This stage determines if the intervention is affordable in the current budget, and if not, what is the additional budget needed to reimburse this new drug or what actions should be undertaken to make it affordable. Some countries do not consider budget impact as they believe it is redundant with the efficiency assessment (point 3), as the ICER threshold is expected to reflect or be adjusted to the affordability. This remains debatable.

Following value assessment, the payers wish to estimate what is the right price for the medicine in question. There are more than 20 pricing methods listed on Wikipedia. In general, four pricing models have been used by the pharmaceutical industry globally:

• Value-based pricing

It sets selling prices on the perceived value to the customer, rather than on the actual cost of the product, the market price, competitors' prices, or the historical price.[*]

• Cost plus pricing

One first calculates the cost of the product, then includes an additional amount to represent profit.[†]

• Willingness to pay pricing

Willingness to pay is the maximum amount a person would be willing to pay, sacrifice, or exchange for a good.[‡]

• Price benchmarking

By observing the quality/value of products of other businesses, a company is able to use price benchmarking to determine a price for their products in relation to where they think they stand amongst the competition (http://www.priceintelligently.com/price-benchmarking).

• Mixed-model pricing

The pharmaceutical industry in the west currently uses a mixed-model that combines value-based pricing and benchmarking.

In the institutionalized health care payer settings, value-based pricing is currently considered to be the most promising model.

For example, in the United Kingdom, there is a growing interest in replacing free-drug pricing with this model:

Decisions about price and guidance will inevitably be controversial, but responsibility for rejecting or restricting access to a new technology will be more appropriately shared between an NHS that cannot afford wider use of an effective but cost ineffective technology and a manufacturer unwilling to accept a price that would make it cost effective (Claxton et al., 2008).

However, using value-based pricing has the following two consequences:

• How to link a value that is perceived to a value that is actually delivered? This is critically important, as at the time of pricing only clinical trial evidence is available and modeling usually carries substantial uncertainty. Rarely, is the actual drug value assessed after launch and confronted with the perceived or expected value at the time of market launch.

• Value is subjective and depends on how customers appreciate value. There are no rules on how to link the value to price except when using the ICER per QALY threshold. However, there are multiple limitations to the use of ICER as the unique measure to set drug prices and a number of countries do not request health economic evidence such as QALY.

In principle, this model assumes that the price of a drug is a function of its relative effectiveness compared with the standard of care. If the product brings about a significantly added value, the manufacturer will get a higher price than the comparator. In cases where the payer cannot afford the

---

[*] Wikipedia, https://en.wikipedia.org/wiki/Value-based_pricing (last modified September 16, 2016).

[†] Wikipedia, https://en.wikipedia.org/wiki/Cost-plus_pricing (last modified September 26, 2016).

[‡] R. Frank and B. Bernanke, *Principles of Macroeconomics*, 5th edition, 2012.

premium price, the financial burden can be shared with other payers (patients or complementary insurers).

IMPORTANCE OF BUDGET IMPACT

In the United States, payers pay for some oncology products $80,0000 to increase life expectancy by 1.2 months. Then by extrapolation, survival of 1 year would be valued at $800,000. In the United States, 550,000 patients die from cancer annually. If new drugs are developed that extend life by 1 year, $440 billion would be needed to purchase this drug. It is obviously unaffordable even for the richest country and unlikely to be affordable for any country. Therefore something will have to change in the way we set prices. It may happen in collaboration with the industry or be imposed on the industry under political pressure with no prior debate.

Therefore, it seems that beyond assessing what is the value (additional health benefit) of a new medicine, we need to be concerned about what is the affordability of the payer to fund this new medicine.

3.5.4 How HTA Evaluation Is Translated into P&R Conditions?

Negative HTA recommendation on the use of a medicine translates into reduced MA in various ways. The impact on price can be through direct reduction of the price by the payer, price–volume agreements, and copayments (Germany). The impact on reimbursement is by reducing the maximum percentage of reimbursement (France).

Further, restrictions can be applied on the scope of prescription of a drug. Partial restriction consists in defining a population of patients or indication, that is narrowed as compared to the marketing authorization of a drug. Full restriction means that a drug will not be included in formularies or in guidelines (Canada and the United Kingdom). Preauthorization of prescription by the payer or a specialist medical centre for a medicine are other means of ensuring that the drug is only prescribed to patients strictly defined by the payer. Finally, MA agreements are contracts between the manufacturers and the payers that aim at obscuring the reallist price of a medicine by allowing often complex discounts or pay-backs on a patient level or by allowing a temporary premium price until stronger evidence over drug's effectiveness or safety is developed. Such agreements are discussed in Chapter 6.

3.6 MARKET ACCESS FOR INDUSTRY

3.6.1 A New Paradigm

Traditionally, obtaining marketing authorization for a new medicine was sufficient for the industry to achieve MA for the product. This involved demonstrating that a new medicine is efficacious, safe, and of good quality (so called *three hurdles*). As health care costs around the world are escalating, new criteria taken in isolation or combined constitute a novel hurdle for the industry, often referred as the *fourth hurdle*. They may include additional benefit, cost-effectiveness, or budgetary impact.

MA activities cover the following three overlapping areas:

- Pricing and reimbursement: Achieve optimal price and full reimbursement
- HTA: Achieve favorable recommendation from the HTA authority
- Formulary: Achieve referencing in various formularies on the hospital, regional (canton, province, Land, etc.) and national (disease-specific NHS formularies) levels

The industry has to create structural organization, governance, and procedures to address those challenges. Specific activities to achieve those new requirements should be integrated in product development plan and life-cycle management. From industrial perspective, MA can be defined in the following ways:

- MA is the process by which a company gets a drug available on the market after regulatory approval, so that it becomes available or affordable to patients.
- MA means managing all the obstacles to achieve the optimal price and unrestricted provision of a medicine to a target population as defined in the marketing authorization.

As such, MA has a very high economical implication and a high strategic and political importance within pharmaceutical organizations. It has become the fundamental element for a product's commercial success and is the driver of all business cases. Unfortunately, company experts in the MA field are rarely involved in informing decision-making. Early integration of MA professionals in the industrial decision-making process guarantees a viable business plan and optimization of return and/or out-licensing potential. Companies that did not achieve access for their product have experienced or will experience a painful setback.

Big pharmaceutical companies have become increasingly focused on this field, which is new and requires multidisciplinary expertise.

3.6.2 Organization in the Pharmaceutical Industry

We performed a review of how pharmaceutical companies are organized to address the MA challenges (Creativ-Ceutical, Paris proprietary research). We identified that there are mainly four models that we named: the dual model, the fragmented model, the integrated model, and the decentralized model. The interesting finding was that there is an important dynamic in the MA organization, and that when products fail to access the market, it is often followed by a deep change in organization. Unfortunately, the important questions that are neither raised nor addressed by the changes are as follows:

- Dual model is a model where the MA activity is split in two main groups. One group often called Health Economics and Outcomes Research (HEOR) or value evidence development is focused on evidence generation reports to the medical department or to the development function. The other is often called Market Access and is more oriented on how to leverage the evidence to gain access, price, and reimbursement and is reporting to the commercial operation or marketing functions. These two groups could constitute two departments or be split among various departments. The HEOR department may be split in health economics (HE), outcomes research (OR), epidemiology, econometrics, and HTA. The MA department may be split in P&R, MA, and HTA. Delineation of the activity varies dramatically from one company to another and is consistently evolving, following power-balance evolution in the company. The strength of this model is to allow a good integration within the development group of the evidence generation group but often leads to divergence of opinion between P&R and evidence generation teams.

- Integrated model is a model where all MA functions are within the same reporting line. Evidence generation group and P&R do have the same leader that may report in various places in the organization. This model allows an integration of the evidence generation and P&R that will join efforts for the same goal. It tends to create a vision gap with the development and marketing teams that may be sometimes difficult to manage.

- Fragmented model has a broad range of departments such as epidemiology, quantitative analysis, HE, OR, MA, and P&R that report in various lines making the collaboration between all those functions complex to manage. There is a trend that the fragmented model produces more evidence and material for supporting the MA, but the outcome assumptions are often not fully consistent.

- Decentralized model is where the various functions operate independently with very little coordination at the global level. Although it tends to disappear, such models still exist, especially in companies that were not exposed to a MA failure. The corporate office leaves this complex topic to be managed regionally, because of the perception that MA has important regional specificities. In this model, the trend is to have good management of local HTA and payer's requirements, with good flexibility for customization. It may be associated to inefficiencies between regions, as some work effort may be duplicated.

The model has to fit the company culture and history, as well as the location of the strength that may support the activity for MA. Each model will have its weaknesses and it is important that those weaknesses are mitigated and managed.

3.6.3 Objective of Market Access Activities

The objective of MA for the industry is to achieve the optimal price with maximum reimbursement level for the approved (in the marketing authorization) population with no burden on prescription or funding procedures. This involves building a value proposition for payers. Importantly, MA activities cover all stages of the drug development process and the value proposition needs to be developed as early as possible. As we have seen above, there are multiple payers and the important challenge for the industry is to build a value proposition in order to satisfy all the stakeholders, consider their different perspectives, and yet remain consistent.

The global objective in the pharmaceutical industry is to increase the return on investment. Because the revenue from a new medicine depends on the product price and sales volume, any increase in the price translates into profit increase. A very small increase in price can have a dramatic effect on the profit. Typically, increasing the price by 1% may increase the profit by 8%. MA has also become the primary driver of volume of sales, even more than marketing. In fact, poor MA will often limit access to a broad patient population, thus limiting the opportunity for volume development. Therefore MA has become the driver of the business case for pharmaceuticals.

SUGGESTED READING

Claxton K. et al. Value based pricing for NHS drugs: An opportunity not to be missed. *BMJ* 2008;336:251.

Creativ-Ceutical, Paris proprietary research.

Garattini L. et al. Pricing and reimbursement of in-patent drugs in seven European countries: A comparative analysis. *Health Pol.* 2007;82:330–339.

http://www.pharmalevers.com/market-access-quiz-.html.

Kristensen FB and Sigmund H (ed.) *Health Technology Assessment Handbook*. Copenhagen, Denmark: Danish Centre for Health Technology Assessment, National Board of Health, 2007.

Market access for Pharma: pulling in the same direction?—a UK perspective from Alan Crofts, http://www.thepharmaletter.com/file/79052/market-access-for-pharma-pulling-in-the-same-direction-a-uk-perspective-from-alan-crofts.html (accessed September 29, 2016).

Market Access—The Definition Depends on the Viewpoint, http://www.paramountcommunication.com/ubc/pdf/01_Market_Access_The_Definition.pdf (accessed September 29, 2016).

Menger C. *Principles of Economics*, Auburn, AL: Ludwig von Mises Institute, 2007.

New Approaches to Gaining Market Access for Pharmaceuticals: Pricing & Reimbursement, Policy Development, and the Role of HTAs. Business Insights, October 1, 2010, http://www.marketresearch.com (accessed September 29, 2016).

Sackett et al. Evidence based medicine: What it is and what it isn't. *BMJ* 1996;312:71.

Surveying, Assessing and Analysing the Pharmaceutical Sector in the 25 EU Member States. Commissioned by European Commission—DG Competition. 2006.

The Global Trade Negotiations Home Page, http://www.cid.harvard.edu/cidtrade/issues/marketaccess.html (accessed September 29, 2016).

4 HTA Decision Analysis Framework

4.1 INTRODUCTION

Rising budget constraints and demands for health care services bring more complexity to the decision process for resource allocation. Innovations and health care progress have been shown to be key drivers of the growth of health care expenditures. In the context of increasing costs for medical care and limited resources, health technology assessment (HTA) was developed as a tool to inform the decision-making, to provide rationalization, and to make the process more transparent. HTA approach consists in evaluating multiple aspects of a new product's value in order to maximize the provided health gain within limited resources.

To answer these questions, HTA must be multidimensional and must analyze both the effects on health (benefits and possible harm) and the consequences in terms of costs of new health interventions in the health care system, in the short and long term. Therefore, HTA relies on evidence-based medicine, as well as on other methodological tools which help in the decision-making process.

The international group for HTA advancement has proposed fifteen key principles to guide the HTA process, embracing four dimensions of HTA: *Structure of HTA programs*, *Methods of HTA*, *Processes for conducting HTA*, and *Use of HTA in decision-making* (Drummond et al., 2008).

4.2 HISTORY

Development of HTA among European countries has been asynchronous and is still ongoing. However, it appears that the implementation of HTA methodology for the evaluation of newly marketed health care interventions is widespread, and HTA is now a mandatory part of the market access process. Battista and Hodge have identified an HTA development pattern in three phases: emergence of HTA, consolidation, and expansion.

Emergence of HTA requires decision-makers to express a demand for information about health care and existing experts and organizations to be able to fulfill this demand. In many cases, this need arises from the dynamics created by one or several *HTA-champions* (academics and/or civil servants) working to create enthusiasm around HTA and promote its development by organizing seminars or short courses. HTA has also been found to generate more demand in countries with a receptive political environment, in favor of health care efficiency (budget constraints, general policy pragmatism, etc.). Emerging HTA focuses on a limited area of health technologies. Technologies with high capital or acquisition costs are concerned first, although pharmaceuticals are excluded from the evaluation. At this stage, HTAs have limited time and budget resources; they are represented by small teams dedicated to isolated questions based on demands of a small group of HTA-enthusiast decision-makers. At this time, HTA results give rise to minimal knowledge translation toward managers, clinicians, patients, and the general public.

HTA then goes through a consolidation phase during which a bigger interest toward HTA develops among the decision-makers. Bigger needs for HTA require the establishment of a priority setting process. The scope of HTA broadens to include more health interventions like pharmaceuticals. The entire care processes are assessed instead of single technologies. This consolidation relies on bigger means dedicated to HTA with the expansion of scientific teams and possible research partnerships with academics or other organizations. Target audience of HTA recommendations expands to more decision-makers, managers, clinicians, patients, and general public.

Once HTA is rooted in the national health care culture, it becomes a part of the official political discourse. During the expansion phase, the demand for HTA continues to grow and diversify. The scope of HTA keeps broadening: existing practices are reviewed as well as new interventions, resources dedicated to HTA increase and knowledge translation to policy or decision-makers, professionals, patients, and general public is more important.

4.3 HTA PROCESSES AND DECISION ANALYSIS FRAMEWORKS

4.3.1 NATIONAL HTA BODIES AND MAIN ASSESSMENT OUTCOMES

With respect to pricing and reimbursement, the United Kingdom, France, Germany, Sweden, and Italy have entered an expansion phase of HTA:

In France, the High Authority on Health (*Haute Autorité de Santé* [HAS]) created in 2004 has incorporated existing HTA organs (*Commission de la transparence*/Transparency Committee [CT] and Commission nationale d'évaluation des dispositifs médicaux et des technologies de santé, Medical Device and Health Technologies Assessment Committee). These bodies evaluate the absolute and the additive therapeutic value of health interventions in order to inform the decision-making organs (Economic Committee of Health Products [CEPS], National Union of the Medical Insurances [UNCAM], and The Ministry of Health). In 2012, a new commission for the pharmaco-economic evaluation of health technologies was created within the HAS: the Economic and Public Health Evaluation Committee (CEESP). It is mandated to perform economic evaluation in order to inform the CEPS of the efficiency of new reimbursed treatments. Within the

HAS, the CT assesses the absolute therapeutic value of health interventions and summarizes it into a medical value score, *Service Médical Rendu* (SMR), on a five-level scale:

- Major or important SMR lead to a 65% statutory health insurance coverage (or 100% for *irreplaceable medicines for serious and disabling conditions*)
- Moderate or weak SMR allows a 30% coverage
- Insufficient medical value forbids coverage by the statutory health insurance

Further, the incremental or added value of health interventions is translated into a medical added value (ASMR) score. It is also a five-level scale ranging from ASMR I *major added therapeutic value* to ASMR V *no added therapeutic value*. The CEESP evaluates the costs and effects of the intervention and determines the opportunity to implement it in the health care system. The results of this pharmaco-economic evaluation and the uncertainty surrounding them are presented in the efficiency notice.

In 1999, in the United Kingdom, the National Institute for Health and Care Excellence (NICE) was mandated by the government to make recommendations on how to best allocate the National Health Service (NHS) budget. NICE has been providing multiple and single technology appraisals, taking into consideration clinical and economic data, in order to enlighten the NHS on the opportunity represented by new health technologies. NICE decisions are officially recognized in Northern Ireland, as well as in England. Wales has its own HTA body, the All Wales Medicines Strategy Group (AWMSG) that has agreed to review all new medicines, not assessed by NICE since 2009. NICE single technology appraisals determine the pharmaco-economic value of health interventions. Main evaluation criterion for the pharmaco-economic evaluation of health interventions by NICE are incremental cost-effectiveness ratios (ICERs), expressed as cost per quality adjusted life year (QALY) and compared to maximum thresholds fixed at £20,000 and £30,000/QALYs.

- ICER<£20,000/QALY—technology is cost-effective
- £20,000/QALY<ICER<£30,000/QALY—acceptability conditional on additional factors
- ICER>£30,000/QALY—a stronger case for supporting the intervention will be required

In Scotland, the Scottish Medicines Consortium (SMC) provides recommendations to the Scottish NHS on all newly licensed medicines, formulations, and indications since 2002. SMC "New Products Assessments" determines whether the new medicine is expected to provide additional health benefits valued by patients at a reasonable net cost for the NHS, or if the new medicine may offer equivalent level of health benefit at equivalent or lower cost for the NHS than its existing comparators. SMC claims, it does not refer to ICER thresholds or willingness-to-pay thresholds, although it mentions that *modifiers* such as orphan drug status, substantial clinical benefits, or absence of other therapeutic options can allow SMC

to accept drugs with more uncertainty in the health economic case or *higher cost per QALY*. SMC also acknowledges referring to the NICE ICER thresholds when making its decisions.

In Germany, products with a marketing authorization are automatically reimbursed by the statutory health insurance. Since 2011, newly marketed products require manufacturers to submit an evaluation dossier to the Federal Joint Committee (G-BA). G-BA then mandates Institute for Quality and Efficiency in Healthcare (*Institut für Qualität und Wirtschaftlichkeit im Gesundheitswesen* [IQWiG]) to examine the dossier and assess the additional benefit or lesser harm provided by the new intervention. There are several levels of additional benefit:

- Major added benefit: a major therapeutic improvement is demonstrated over an appropriate alternative, such as a significant reduction of symptoms, a significant improvement in overall survival, and/or the avoidance of serious relevant side effects.
- Significant added benefit: a significant therapeutic improvement is demonstrated over an appropriate alternative, such as a significant reduction of symptoms, a moderate improvement in overall survival, and/or the avoidance of serious relevant side effects.
- Small added benefit: a slight (but not minor) therapeutic improvement is demonstrated over an appropriate alternative, such as a reduction of nonsevere symptoms or prevention of a relevant side effect.
- Unquantifiable added benefit because of irrelevance or lack of scientific data.
- No added benefit.
- Benefits are inferior to those of the alternative therapy.

The evaluation performed by IQWiG assesses the marginal therapeutic value of the products. Interventions with an additional benefit must go through a pharmaco-economic assessment performed by IQWiG in order to determine the cost-effectiveness of the intervention. ICERs of these interventions are used to build an efficiency frontier. Cost-effectiveness of the new intervention is appreciated by comparing its position relatively to this efficiency frontier.

In Sweden, the Dental and Pharmaceutical Benefits Agency (*Tandvårds-och läkemedelsförmånsverket* [TLV]) evaluations assess both the additional therapeutic value of health products and the efficiency they may generate. The assessment is mainly based on pharmaco-economic evaluation.

In Italy, the Agenzia Italiana del Farmaco (Italian medicines agency [AIFA]) is the main authority for pharmaceuticals. Two AIFA commissions, the Technical Scientific Committee (CTS) and the Pricing and Reimbursement Committee (CPR), assess new interventions. AIFA's assessments result in the characterization of the product's degree of innovation (important, moderate, and mild) based on a therapeutic algorithm, assessing the availability of comparators and the benefits brought by the product toward these comparators, as well as other descriptive and comparative clinical and economic data.

Spain is still in a consolidating phase of HTA at the national level: national pricing and reimbursement procedures are simultaneous. Treatment evaluations are performed by the General Directorate for Pharmacy and Medical Devices (*Dirección General de Farmacia y Productos Sanitarios* [DGFPS]), a part of the central ministry of health. Submission dossier is assessed by DGFPS leading to both reimbursement and pricing recommendations. HTA in Spain is not transparent. Pharmaceutical companies have been complaining about such issues since the economic crisis of 2008 and the implementation of significant cost-containment measures.

More country-specific details are presented in Chapters 12 through 21.

4.3.2 DECISIONS IMPACTED BY THE ASSESSED OUTCOMES

Although the same products' characteristics can be assessed by the national HTA agencies, they are not determined similarly and do not impact decisions in the same way. Furthermore, the scope of the assessment varies between countries and agencies as described below.

- Absolute therapeutic value:
  - It is assessed by the Transparency Committee in France. The SMR determines whether the intervention should be *reimbursed* and what level of coverage should be provided by the statutory health insurance.
  - In Spain, DGFPS considers the product's absolute therapeutic value in consideration with the *reimbursement* recommendation.
- Additional/marginal therapeutic value:
  - In France, the CT translates the added therapeutic value into an ASMR level used during the *pricing* negotiations between the CEPS and the manufacturer.
  - In Germany, the additional benefit determined by IQWiG establishes whether the manufacturer can negotiate a reimbursed *price* with the statutory health insurance, and how much higher it can be, as compared to the existing products.
  - In Sweden, the interventions with a marginal benefit (*no other available medicines are significantly more suitable*) can be *reimbursed* within the pharmaceutical benefits scheme. The higher the marginal benefit is, the higher a reasonable *price* can be set.
  - In Italy, the degree of innovation helps determine the inclusion of the product on the national *reimbursement* list and the reimbursement setting (both in-patient and out-patient, or in-patient only). The CPR then leads the *pricing* negotiations with the manufacturer and the Committee for Economic Planning (CIPE).
- Budget impact of the intervention in the health care system is a widely assessed parameter. Although it is sometimes not considered explicitly in the making of the recommendation, budget impact analyses are referred to during the negotiation processes between retailers and decision-making bodies.
  - In Spain, budget impact analyses (BIA) are considered in the making of both *reimbursement* and *pricing* recommendations.
- Cost-effectiveness:
  - In France, Sweden, and Germany, the cost-effectiveness of the health care technologies is used in the *pricing* negotiations.
  - NICE technology appraisals based on cost-effectiveness evaluation lead to the *implementation* of the recommended technologies within the NHS in England, Wales, and Northern Ireland. Similarly, SMC's new products assessment determine whether the technology will be recommended in the NHS Scotland.
  - In Italy, cost-effectiveness data are considered for both *pricing* and *reimbursement* decisions. Similarly, in Spain, cost-effectiveness analyses are expected to substitute BIA in the DGFPS recommendation process.
- Retailer costs, planned profits, prices in other countries, and price of alternative interventions:
  - These parameters are considered in the *pricing* recommendations made in Spain.
- Equity, is a criterion in HTA conducted by, for example, NICE.

4.3.3 HTA CORE MODEL® EUnetHTA

European network for Health Technology Assessment (EUnetHTA) has developed a 9-domain Core Model® for HTA that includes health problem and the current use of technology, description and technical characteristics of technology, safety, clinical effectiveness, costs and economic evaluation, ethical analysis, organizational aspects, social aspects, and legal aspects (http://www.eunethta.eu/hta-core-model).

4.3.4 EVALUATION CRITERIA AND PROCESSES FOR HTA AND PRICING

Although there is an overall trend to comply with the good practice among HTA organizations (e.g., EUnetHTA guidelines and the HTA key principles described by Drummond et al., 2008), evaluation of the parameters of interest can differ across countries depending on the decisions impacted by the evaluation and the structure of the health care system. Further, the differences in the scope of the evaluation determine the type and the sources of data to be assessed, as well as the perspective used for economic evaluation.

4.3.4.1 HTA

All new original drugs are assessed by HTA bodies in France, Germany, Italy, Spain, and Sweden. However, in Germany, hospital-only medicines are not subject to evaluation by HTA bodies. Generics are assessed in Italy, Spain, and Sweden, but in the case of generics the submission of health economics data is not required.

In England, NICE assesses only new drugs identified through specific criteria, such as patient clinical benefit, public health interest, potential budget impact to the NHS (excluding vaccines and HIV products).

We define a formal HTA process as a process where clear terms and objectives are set, there is a transparent decision framework process, a meeting agenda is available, and decisions are publicly available and based on the assessment of evidence submitted by the manufacturer. The process of HTA is formal in France, Germany, Sweden, and the United Kingdom. In contrast, the process at the national level is informal in Italy and Spain, and no decision reports are published in these countries.

We consider HTA as *ex ante*, when HTA precedes price negotiation or reimbursement decision. In that case the HTA enlightens the price negotiation from the payer perspective. An *ex ante* HTA process is adopted in France, Germany, Italy (national level), and Spain (national level). These countries are also called *price giver* countries, because ultimately the price is given by the state representatives leading the negotiation.

An *ex post* process is adopted in Germany, Italy (regional level), Spain (regional level), Sweden, and United Kingdom. The price is set on manufacturer's proposal and authorities will either accept it or reject it, if considered inappropriate. These countries are also called *price takers* as they take or reject a price offer from the manufacturer.

The German situation is quite unique as the country is initially a price taker as the manufacturer sets the original price and it is accepted with no negotiation (free pricing). Then the HTA is performed and a price negotiation is engaged after HTA assessment (price giver).

In France, HTA decision criteria to assess the benefit over the next best alternative are primarily based on absolute therapeutic difference in efficacy or safety and on cost effectiveness for innovative products (with less than 5% of products undergoing HTA). In Germany, the decision is based on relative efficacy or safety. The relative efficacy is assessed through the 95% upper limit of confidence interval of the relative risk.

In Italy and Spain, absolute, relative therapeutic values, as well as drug budget impact are used as main decision criteria.

TLV in Sweden uses absolute, relative therapeutic value, and systematically also cost-effectiveness that is ultimately the main criterion.

The decision of NICE in the United Kingdom is primarily based on cost-effectiveness analysis.

4.3.4.2 Pricing

In France, a national reference pricing by active substance or therapeutic class is used to define the price for reimbursed drugs. Exceptions to this rule are rare and include the use of international reference pricing in the case of innovative drugs. However, confidential discounts from the list prices can be requested and are negotiated with the pricing committee. List price negotiations are performed in France for all other products, except hospital products.

In Germany, different pricing rules are as follows:

- Free pricing: for drugs eligible for early benefit assessment (EBA): up to 12 months after launch
- For drugs noneligible for EBA: national reference pricing by active substance, pharmacological class, and therapeutic class
- Price negotiation: drugs eligible for EBA with acknowledged added benefit or with no reference price groups and no option for creating a price reference group
- International reference pricing: considered as supportive criteria, especially in case of price arbitration

In Spain and Italy, national reference pricing by active substance is used in addition to price negotiations and international reference pricing as supportive criteria.

In the United Kingdom, free pricing is adopted with indirect profit control through the Pharmaceutical Price Regulation Scheme (PPRS).

All the countries above except Germany also use managed entry agreements (also known as market access agreements) to define net drugs prices that are confidential (not disclosed). The list price is also called facial price as it is the visible one whereas the actual price remains confidential.

4.3.4.3 Market Access Agreements

Market access agreements (MAAs) are specific agreements between payers and drug manufactures that specify the terms and conditions under which a product will get access to the market. They are described in detail in Chapter 6. MAAs can be classified into two distinct groups: financial agreements concerned with the financial aspect of the transaction (e.g., price volume agreement, discount or rebate, cap price, or cost sharing) and outcome-based agreements concerned with real-life outcomes or usage of the new drug (e.g., payment for performance [P4P] and coverage with evidence development [CED] agreements). Here, we present a comparison of popularity of different kinds of MAAs among selected European countries (Table 4.1).

TABLE 4.1

Popularity of Various Types of Market Access Agreements for Medicines in Selected European Countries

	Price-Volume Agreement	Cap Volume/ Dose	Price Discount	P4P	CED
France	✓✓✓	✓	✓✓✓	✗	✓✓
Germany	✓	✗	✓✓✓✓	✓	✗
Italy	✓✓	✓✓✓✓	✓✓✓	✓✓✓✓	✗
Spain	✓✓	✓✓	✓✓✓	✓✓	✓
Sweden	✓	✗	✗	✓	✓✓✓✓
UK	✗	✓✓	✓✓✓✓	✗	✓

Notes: P4P: Payment for performance; CED: Coverage with evidence development; x: very rare/not used.

4.3.5 Value Assessment Frameworks

Because prices of drugs, especially in oncology, are increasingly high, they challenge the sustainability of national health care systems in Europe and can lead to catastrophic out-of-pocket patient copayments in the United States. As the 'value for money' is an abstract concept, a number of initiatives emerged to provide a more objective assessment of drug value. The so-called value assessment frameworks (VAFs) are an emerging approach to grade the additional value of a drug compared to a reference product.

The objective of VAFs is to provide patients, physicians, and payers with a tool to support decision making. To date, most developed countries have a formalized decision-making framework to appreciate drug value via the HTA process. However, the link between the clinical benefit and the value remains a deliberative process in most national HTA agencies. Only Germany has developed a very clear framework that allows one to score the drug value in terms of additional benefit over its comparator. VAF tools aim to minimize the subjectivity of the deliberative process during the appraisal and offer a transparent, reproducible assessment of additional benefit.

Examples of VAFs include the following:

Oncology:
1. The Conceptual Framework to Assess the Value of Cancer Treatment Options, developed by the American Society of Clinical Oncology (ASCO)
2. Memorial Sloan Kettering Cancer Center's DrugAbacus
3. The National Comprehensive Cancer Network (NCCN) Evidence Blocks™
4. The European Society for Medical Oncology Magnitude of Clinical Benefit Scale (ESMO-MCBS)

Cardiovascular:
1. The American College of Cardiology and the American Heart Association (ACC-AHA) Statement on Cost/Value Methodology in Clinical Practice Guidelines and Performance Measures

All conditions:
1. The Institute for Clinical and Economic Review (ICER) Value Framework

Currently, dominant frameworks are those of ASCO and ESMO, followed by ICER and DrugAbacus. Although VAFs are still an emerging concept, they are expected to develop and become increasingly accurate. Rather than relying on subjective judgment of HTA committee members, HTA agencies will be able to use the objective VAF scores or grades to underpin their value assessment decisions. HTA agencies worldwide have not adopted VAFs yet, but it is very probable they will do so in the near future.

4.4 FRANCE

4.4.1 SMR

In France, SMR determines the level of coverage of the intervention by the statutory health insurance. It appreciates the treatment's *absolute* efficacy, safety, and public health interest based on the following criteria:

- Descriptive benefit/risk ratio of the treatment assessment:
  - Benefits assessed include: survival, reduced disability, and surrogate outcomes with consideration for the limitations of the results and the bias toward effectiveness. Public health interest of the interventions constitutes another criterion. Randomized controlled trials and observational studies can be considered
  - Risks embrace treatment emergent adverse events and additional morbidity-mortality due to the intervention. All available safety data should be considered
- Disease burden: epidemiology, consequences for the patients, and health care system
- Target population: definition of the target population often based on the marketing authorization, size of the target population must be determined as well as the distributions of dosage and treatment duration among it
- Preventive, curative, or symptomatic nature of the treatment

SMR assessment is purely descriptive and does not take into consideration the comparators. SMR is assessed considering the statutory health insurance perspective.

4.4.2 ASMR

ASMR assesses the therapeutic innovation brought by the intervention in the same setting as the SMR (indication and target population). It aims to determine the added benefit of the intervention based on the following:

- A descriptive benefit/risk assessment of the considered intervention and its appropriate comparator(s)
- Psychological, social, and ethical aspects of the compared interventions
- Organizational and professional concerns such as other interventions linked to the assessed technology or its place among the broader treatment strategy

Even though it is a comparative process, it is based on descriptive outcomes (no differential/incremental results), from the statutory health insurance perspective, considering every clinically relevant comparators available (drugs, devices, procedures, etc.).

Newly marketed health products for which the manufacturer has required an inscription on the reimbursement list(s) are assessed by the CT that gives a verdict on the therapeutic absolute and added values of the intervention after review of the dossier submitted by the manufacturer. Once the dossier has been submitted, the recommendation should be issued after a maximum of 180 days (according to the European transparency directive).

The preliminary version of the recommendation is sent to the manufacturer in order to allow them to answer the committee's questions, add comments, or request a hearing with CT if required.

ASMR and SMR are reviewed every 5 years or sooner on request.

4.4.3 EFFICIENCY NOTICE

To determine the cost and effectiveness of health interventions, the CEESP favors cost utility analyses (CUA) or cost-effectiveness analyses (CEA) if CUA is inappropriate. ICERs then determine which products are dominated or dominant. According to the type of identified health effects, final outcomes of these analyses should be expressed in QALYs in CUA or in life-years gained in CEA. To determine costs and effectiveness of the interventions, modeling is the preferred approach.

- Effectiveness outcomes consist of comparing QALYs in CUA and survival in CEA.
- Compliance and acceptance concerns can impact efficacy and safety through modeling and can be considered through the use of health related quality of life (HRQoL).
- Patients' satisfaction and preferences are integrated through HRQoL in CUA.
- Only direct costs are compared. Resource utilizations brought by recourse to health interventions are associated to the production costs of these resources. Distribution of economic burden among payers (statutory health insurance, private insurance, and patients) should be identified.
- Organizational and professional considerations are taken into account.
- Uncertainty around CEESP's recommendation is assessed based on sensitivity analyses: probabilistic sensitivity analyses should be performed, as well as univariate and multivariate deterministic sensitivity analyses. Data uncertainty with respect to the methodological guidelines are also taken into account to assess uncertainty around results and recommendations of the economic evaluation.

CEESP assesses the cost-effectiveness of innovative health technologies (ASMR level I to III), candidates for reimbursement, that are likely to significantly impact the expenses of the public health insurance (interventions expected to generate more than €20,000,000 annual sales revenue during the second full year of exploitation). CT and CEESP assessments are conducted simultaneously.

Manufacturer can initiate an early contact procedure with HAS that will lead to the elaboration of a preliminary version of the economic assessment dossier. Manufacturers are interviewed after the examination of the dossier, to answer the requests or comments of the commission, to give opinions, and to provide supplementary data.

All relevant interventions should be compared to the assessed one, using an *all-payers* perspective to identify all benefits and all costs generated by the intervention and who they concern: patients, payers, or providers.

4.5 ENGLAND, IRELAND, AND WALES—HEALTH TECHNOLOGY APPRAISAL

ICERs of assessed interventions and comparators are determined through modeling:

- Effectiveness and safety are determined through changes in HRQoL: treatment efficacy and adverse reactions act on survival and HRQoL that are translated into QALYs. These data should be extracted from randomized controlled trials (RCTs), although nonrandomized studies may be accepted or required. Study results are synthesized in meta-analyses. Network meta-analysis of pair-wise head to head RCTs can be performed if required.
- Patient's compliance and acceptance of treatment can impact efficacy and tolerance that will have consequences on survival and quality of life.
- Patients' satisfaction and preferences can be integrated in HRQoL data.
- The considered costs are those related to resources under the control of the NHS and Prescribed Specialised Services (PSS) and are measured correspondingly.
- Uncertainty should be assessed using PSA with results presented as confidence ellipses and scatter plots on the cost-effectiveness (CE) plane, acceptability curves relatively to ICER thresholds, and tabulated results. Uncertainty around the results originating from data or modeling is considered.

NICE's appraisal is referred to in the following cases:

- The technology is likely to have a significant health benefit.
- It is likely to have a significant impact on other health-related government policies (e.g., reduction in health inequalities).
- The technology is likely to have a significant impact on NHS resources.
- There is significant inappropriate variation in the use of the technology among providers or regions.
- The institute is likely to add value by issuing a national guidance.

Manufacturers provide the evidence; they can answer to questions and comments on a draft recommendation. Patients, health professionals, as well as the general public are involved in the review process and committee participation.

All relevant comparators should be considered in the appraisal process, considering a NHS and PSS perspective.

4.6 SCOTLAND—NEW PRODUCT ASSESSMENT

SMC methods to produce new product assessments are very similar to the NICE methods for single technology appraisal, although SMC acknowledges that cost-utility methodology might not be the most appropriate and considers cost-effectiveness study submission in such cases.

SMC secretariat leads a *horizon scanning* process enabling them to identify new technologies during the marketing authorization process and inquire manufacturers about specific data and other submission requirements, as well as timing advice.

Submission is first reviewed by the SMC New Drug Committee (NDC) that reports evidence-based data from the submission to the broader SMC board in charge of the assessment, which can adopt a broader perspective on the matter.

Evaluation process is very transparent. It implies senior NHS managers, public representatives, and the pharmaceutical industry. Patient access schemes, consisting of negotiated rebates, may be proposed by pharmaceutical companies to improve the cost-effectiveness of their products.

4.7 GERMANY—ADDITIONAL BENEFIT AND COST-BENEFIT

The main evaluation criterion is the existence of an additional benefit brought by the assessed intervention. Evaluation of the additional benefit and harm is based on the following three main patient relevant outcomes:

- Mortality (intervention's ability to increase patient's life expectancy)
- Morbidity (improvement in patient's health state, reduction of disease duration, burden, and variation)
- HRQoL

Supplementary outcomes may precise the benefit but cannot constitute an additional benefit by themselves (time and effort invested in disease and intervention by patients, relatives and carers, treatment satisfaction of patients, etc.).

Benefit assessment should be based on RCTs and mixed treatment comparison of RCTs to demonstrate certainty of the results. Surrogate patient outcomes should be validated if used.

Uncertainty around every identified additional benefit or lesser harm is assessed using the level of proof of the results' sources as follows:

- Randomized studies with a low risk of bias constitute high certainty of results
- Randomized study with a high risk of bias give results with a moderate certainty
- Results from a nonrandomized comparative study have a low level of certainty

Conclusions on the level of evidence of the additional benefit or lesser harm depend on the number of sources and the level of certainty of the results:

- Proof of additional benefit/lesser harm
- Indication of additional benefit/lesser harm
- Suggestion of additional benefit/lesser harm

Interventions with an additional benefit must go through a pharmaco-economic assessment performed by IQWiG. The intervention's effect (benefits and risks) used in this analysis are based on the additional benefit determined previously, in the benefit assessment. Study data can be used to model effectiveness and events occurring during the course of the disease. IQWiG considers the use of many methodologies (cost-effectiveness, cost-utility, cost-benefit analyses [CBA], etc.).

Benefit should be quantified using survival or disease adjusted life years in CEA (weighted with HRQoL data in CUA or monetized in CBA). Other effectiveness outcomes could be used if necessary.

Direct medical costs and some direct nonmedical costs if appropriate should then be included. Indirect costs are not to be considered. Intervention costs due to extra-survival should be considered.

IQWiG is commissioned by the G-BA or the Federal Ministry of Health to perform HTA for new health interventions (*early benefit assessment*), based on the manufacturer's dossier submitted to the G-BA.

During the process of the assessment, IQWiG involves medical/health professional experts and patient representatives. Furthermore, IQWiG organizes online hearing procedures opened to participation of the general public.

The assessment is made considering all relevant therapeutic alternatives and a statutory health insurance's and patients' perspective (SHI reimbursement + patient's copayments).

4.8 SWEDEN—MARGINAL BENEFIT AND COST-EFFECTIVENESS

TLV aims to identify the cost-effectiveness (cost-utility) and marginal therapeutic benefit of the intervention.

To produce evidence on the therapeutic benefit, TLV bases its evaluation on pivotal phase 2 clinical trials, phase 3 studies, and indirect comparative studies if no direct comparative study is available. Additional benefit must be proven compared to medicines used in current practice.

Cost-effectiveness data should be produced following the general guidelines for economic evaluations of the Pharmaceutical Benefits Board (LFN). The guidelines state that costs and effects of health interventions should be determined using a societal perspective. In this analysis, the most appropriate comparator should be used: the most commonly used in Sweden, whether it is a drug, another kind of treatment or no treatment. Modeling is used to extrapolate efficacy data to effectiveness, adjusting foreign data to the Swedish setting, resource consumption and costs.

TLV evaluates new medicines on reception of an application from the marketing company. The LFN within TLV makes the decisions on the coverage and prices of new medicines.

If the applying company does not agree with the decision, it may appeal to the general administrative court.

After the company has sent its application for pricing and subsidies, a meeting can be set up to allow both the company and TLV to ask the questions they may have about the application.

4.9 ITALY—DEGREE OF INNOVATION AND COST-EFFECTIVENESS

Prior to evaluation process, AIFA categorizes the intervention based on their therapeutic target:

- Products that are used for the treatment of serious diseases that may cause death, hospitalization, or permanent disability, such as HIV, Parkinson's disease, and cancer
- Products for the treatment of risk factors for serious diseases (hypertension, obesity, osteoporosis, etc.)
- Drugs for the treatment of nonserious diseases

The degree of innovation is then assessed based on the following *therapeutic algorithm*:

1. Availability of existing treatments in the therapy area covered by the following intervention:
   a. Drugs treating patient population for which no alternative effective treatment exists
   b. Drugs aimed at diseases with subgroups of treatment-resistant patients or nonresponders to the first-line therapy (e.g., anti-HIV, anti-cancer drugs)
   c. Drugs aimed at diseases for which alternative treatments exist. In this case, the innovation is assessed comparative to existing treatments:
      i. Drugs with a better efficacy or safety or with a more advantageous pharmacokinetic profile than the existing alternatives
      ii. Drugs presenting a pharmacological innovation (new mode of action) but no improvement over existing therapies
      iii. Drugs with a technological innovation (new molecule) but no improvement over existing therapies
2. Additional benefit of the new intervention as well as the level of evidence are characterized:
   a. Major benefit demonstrated on clinical endpoints or on validated surrogate end points
   b. Partial benefit: improved or minor benefit demonstrated on clinical or validated surrogate end points or major benefit with limited evidence
   c. Minor or temporary improvement in some disease symptoms

These data are then aggregated into a degree of innovation (important, moderate, and mild).

The information taken into account by the committees when making pricing and reimbursement decisions includes the following:

- Information provided by the manufacturer, including market registration, safety data, and proposed price
- Consumption and expenditure data provided by the Medicines Utilization Monitoring Centre (OsMed)
- Product's therapeutic characteristics (including its relative value compared with established products)
- Disease-specific criteria (severity of illness, size of potential target population, and special medical needs)
- Production methods and costs
- Results of clinical trials
- Cost-effectiveness analyses
- Risk-benefit studies comparing the product to existing medicines
- Daily treatment cost/cost of therapy versus comparative products

Pharmaco-economic evaluations are more and more provided by manufacturers. Innovative products offering therapeutic benefits over established treatments can in particular benefit from such studies, given that evidence of cost-effectiveness can impact pricing negotiations. Although guidelines for economic evaluations have been published by the Italian group of pharmaco-economic studies, there are no official guidelines to conduct health economic evaluations in Italy.

Assessment process in Italy is not very transparent: participation and consultation of stakeholders are very limited, as is the transparency of the decision-making process. The decision-making process is strongly influenced by the AIFA's general director. This results in a hierarchic, non-participative organization, with many cost-containment priorities, and may lead to low consideration of the industrial perspective.

4.10 SPAIN—REIMBURSEMENT AND PRICING RECOMMENDATION

National pricing and reimbursement procedures are simultaneous in Spain. Once a new product has been granted a marketing authorization, the manufacturer must provide the DGFPS, a pricing and reimbursement application that will lead to two recommendations.

The two recommendation processes do not consider the same parameters:

- Reimbursement recommendation is made based on:
  - Absolute therapeutic value of the product assessed, considering the severity, duration, and consequences of the condition treated, the existence of a clinical need, the therapeutic, and social value of the product.
  - The level of innovation of the technology determines how the costs assessed through a budget impact or cost-effectiveness analysis compare to

therapeutic alternatives (external reference pricing or internal reference pricing).

The recommendation is then forwarded to the National Commission for the Rational Use of Medicines (*Comisión Nacional para el Uso Racional del Medicamento* [CNURM]) responsible for the reimbursement decision.

- Pricing recommendation assesses the following:
  - Total cost of the product from the manufacturer's point of view (production, distribution, sale costs). Research and development costs are considered as well.
  - Manufacturer's planned profits and 3-year sales forecasts, as well as budget impact or cost-effectiveness analysis.
  - Manufacturer's net price to wholesalers in the firm's home country and other European countries (France and Italy in particular) and net prices to wholesalers of similar drugs in Spain and other countries.

The recommendation is sent to the Inter-Ministerial Pricing Commission (*Comisión Interministerial de Precios de los Medicamentos* [CIPM]), the Ministry of health organ in charge of the pricing negotiations during which clinical data and innovation concerns can be considered.

HTA in Spain is supposed to evolve in the upcoming years. Local HTA agencies have developed alongside the National Institute for Health Carlos III (ISCIII, Insituto de Salud Carlos III). All of these HTA organs are supposed to be integrated into a national HTA organ, whose authority and scope of activities remain to be decided.

SUGGESTED READING

Battista RN, Hodge MJ. The "natural history" of health technology assessment. *Int J Technol Assess Healthcare*. 2009;25(Suppl 1): 281–284.

Drummond MF, Schwartz JS, Jonsson B, Luce BR, Neumann PJ, Siebert U, et al. Key principles for the improved conduct of health technology assessments for resource allocation decisions. *Int J Technol Assess Healthcare*. 2008;24(3):244–258; discussion 362–368.

Goetghebeur MM, Wagner M, Khoury H, Levitt RJ, Erickson LJ, Rindress D. Bridging health technology assessment (HTA) and efficient health care decision making with multicriteria decision analysis (MCDA): Applying the EVIDEM framework to medicines appraisal. *Med Decis Making*. 2012;32(2):376–388.

http://www.inahta.org/hta-tools-resources/
http://www.ispor.org/HTARoadmaps/Default.asp
http://www.eunethta.eu/hta-core-model

5 Early HTA Advice

5.1 OVERVIEW OF THE EARLY HTA ADVICE PATHWAYS

Medicine developers have an opportunity to gain feedback from regulators and HTA bodies, early in the development process of a medicine. The authorities concerned with these schemes use various terms, such as *early dialogue* or *scientific advice*; however these are often specific to an organization and there is no consensus on terminology. In this chapter, we use the general term *early HTA advice* or the organization-specific terms when relevant.

Such advice can help pharmaceutical companies establish what evidence the HTA authorities will need in order to determine a medicine's benefit-risk balance (in the marketing authorization process) and its *value-for-money* in real-life use (in the HTA process). Companies may seek several kinds of early HTA advice in different stages of the drug development process, as described in the following sections and summarized in Table 5.1.

5.1.1 HTA-EMA PARALLEL SCIENTIFIC ADVICE

Manufacturers can apply for parallel scientific advice from European Medicines Agency (EMA) and national HTA bodies at any stage of development of a medicine, whether the medicine is eligible or not for the centralized authorization procedure in the European Union.[1,2]

In the process, EMA and HTA bodies are equal partners. Manufacturers can be flexible in the choice of HTA bodies and the EMA can facilitate the contacts. If considering more than five HTA bodies, additional discussion with an EMA scientific advice officer is recommended. Typically, companies engage early in informal discussions with HTA bodies and EMA announcing intention for procedure, product, and timescale, and which HTA bodies will participate. More details on how to initiate the various early advice pathways is presented below.

5.1.2 MULTIHTA ADVICE

The SEED consortium (Shaping European Early Dialogues) was a European commission project led by the *French Haute Autorité de Santé* (HAS) in years 2013–2015.[3] It consisted of 14 national and regional HTA bodies.[4] The SEED consortium developed a proposal for a permanent model for early dialogue that would govern the process from 2016. The permanent process has not been established at the time of writing this publication.

5.1.3 EUNETHTA PILOT ASSESSMENT OF RELATIVE EFFECTIVENESS

Pilot assessment of relative effectiveness was a project of the European network for Health Technology Assessment (EUnetHTA) which aimed to test methodology, procedures, and national or local implementation of joint rapid relative effectiveness assessments (Rapid REA), and to test the capacity of national HTA bodies to collaborate and produce structured rapid core HTA information on relative effectiveness.[5] As a result of the project, methodological guidelines for Rapid REA of pharmaceuticals were published by EUnetHTA in March 2013.[6] The SEED project described above builds on the results of the EUnetHTA pilot project.

5.1.4 ADAPTIVE PATHWAY

This is an accelerated scientific advice pathway of EMA for therapies indicated for serious conditions with high unmet needs.[7] It requires that there is an iterative development with use of real-life data. It provides the possibility to engage various stakeholders including regulators, HTA bodies, and patient representatives in multiple discussions along the development pathway.

5.1.5 PRIORITY MEDICINES SCHEME

EMA has developed a priority medicine scheme (PRIME) to optimize the development and accelerated assessment of medicines of major public health interests.[8] PRIME reinforces early dialogue and builds on regulatory processes such as scientific advice to optimize the generation of robust data and the accelerated assessment procedure to improve timely access for patients to priority medicines.[9] This kind of early advice focuses on key development milestones, with the potential involvement of multiple stakeholders including HTA bodies and payers, as well as patient organizations, where relevant.

5.2 NATIONAL EARLY HTA ADVICE PROGRAMS

Several countries have put in place early HTA advice programs for the clinical development plan of medicines. These programs are described below.

5.2.1 FRANCE

HAS offers prospective advice for companies. It can be conducted in English. It is confidential, not legally binding, and there are no fees involved.

TABLE 5.1
Types of Early HTA Advice Offered by EMA and HTA Bodies

Type of Advice (Organization)	Prerequirements	Time Frame (Days)	Procedure Details Including Timelines (and Required Documents)	Fees	Specificities of the Process (Choice of HTA Bodies, Nonbinding, Key Focus of Questions, etc.)
HTA–EMA parallel scientific advice (EMA and chosen national HTA bodies)	At any stage of development of a medicine, whether the medicine is eligible for the centralized authorization procedure or not. This may include postauthorization safety and efficacy studies and risk management planning incorporating risk minimization measures Very early with nonclinical proof of concept and no clinical data When exploratory clinical data are available	45–80	Companies engage early in informal discussions with HTA bodies and EMA announcing intention for procedure, product and timescale, and which HTA bodies will participate. **Premeeting. Day 0–D59** • Assessment of the briefing package (book) by EMA (list of issues sent to applicants) and HTA bodies • Presentation for the face-to-face meeting to be sent by the applicant within 2 weeks of receipt of the list of issues to EMA and HTA bodies, with written responses, if requested • EMA will arrange a closed preparatory teleconference with HTAs after responses to the list of issues and presentation, to identify critical divergences between EMA/HTAs (communicated to the applicant in advance) **Meeting. Day 60–62** • A face-to-face meeting between all stakeholders, lasting approximately 4 hours (EMA premises) **Outcome. Day 63–70** • Minutes of the meeting are circulated by the company within 5 days to all participants • EMA final advice letter contains Committee for Medicinal Products for Human Use (CHMP) regulatory advice only • HTA bodies feedback is provided directly to company during the face-to-face meeting, or by annotating the applicant's minutes, or by providing written answers	EMA charges €63,000–84,000 for initial request and €32,000–42,000 for follow-up to the initial request. Fee reduction for small and medium-sized enterprises and for orphan drugs. Participating HTA agencies may also charge a fee.	• EMA and HTA bodies are equal partners • Flexible in choice of HTA bodies • EMA can facilitate contacts • HTA bodies are chosen by the applicants • Not legally binding • Confidential

(Continued)

TABLE 5.1 (Continued)
Types of Early HTA Advice Offered by EMA and HTA Bodies

Type of Advice (Organization)	Prerequirements	Time Frame (Days)	Procedure Details Including Timelines (and Required Documents)	Fees	Specificities of the Process (Choice of HTA Bodies, Nonbinding, Key Focus of Questions, etc.)
MultiHTA advice (A consortium of national HTA bodies led by the French health agency HAS)	• Generic or biosimilar products are out of scope • It focuses on development strategies and not on preassessment of data. • The advice is prospective; advice on on-going pivotal trials will not be accepted. • For drugs, it should ideally be requested during the phase II to discuss the content of the planned phase III, that is, planned confirmatory trial(s) and the economic rationale.	110	Intent letter. 4 months before meeting Briefing book submission. Day -90 Upgradation of the briefing book. Days -75 to -7 exchanges via e-mail between applicant and HTA bodies HTA Bodies release written positions. Day -7 Early advice meeting. Day 0 1. Preliminary discussion among HTA bodies only 2. Face-to-face meeting of HTA bodies with the company 3. Conclusions among HTA bodies only **Minutes. Day 10** The company provides the draft detailed minutes of the meeting. **Minutes revision. Day 20** The draft minutes are revised in writing by participating HTA bodies each correcting only the position of their agency and commonly agreed statements.	To be announced in 2016	• Early advices are restricted to one indication; however one or more lines of treatment may be discussed within this indication. • Questions should be related to HTA in the view of reimbursement and pertaining mainly to relative effectiveness, economic aspects, and other areas relevant for reimbursement. • The company can choose areas to be discussed • Not legally binding • Confidential
EUnetHTA pilot assessment of relative effectiveness (EUnetHTA and member HTA bodies)	Not taking new applications. To be replaced by the SEED permanent process from 2016.				

(Continued)

TABLE 5.1 (*Continued*)
Types of Early HTA Advice Offered by EMA and HTA Bodies

Type of Advice (Organization)	Prerequirements	Time Frame (Days)	Procedure Details Including Timelines (and Required Documents)	Fees	Specificities of the Process (Choice of HTA Bodies, Nonbinding, Key Focus of Questions, etc.)
Priority medicines scheme *PRIME* (EMA and chosen national HTA bodies)	• Medicines that may offer new therapeutic options to patients who currently have no treatment options, or a major therapeutic advantage over existing treatments. • The medicines would have to show preliminary clinical evidence indicating that it has the potential to bring significant benefits to patients with unmet medical needs and hence be of major interest from a public health and therapeutic innovation perspective. • EMA proposes earlier entry into the scheme for micro-, small-, and medium-sized enterprises (SMEs) and applicants from the academic sector on the basis of compelling nonclinical data and tolerability data in initial clinical trials.	40	The applicant should submit a request for PRIME support electronically to EMA including a justification and summary of available data. Eligibility submissions are accepted according to a schedule published at http://www.ema.europa.eu/ema/index.jsp%3Fcurl%3Dpages/regulation/general/general_content_000660.jsp%26mid%3DWC0b01ac058096f643 Upon receipt of the request, one Scientific Advice Working Party (SAWP) reviewer and one EMA scientific officer will be appointed for the procedure to start in accordance with published timetables, as follows: Day 1 Start of procedure (SAWP 1 meeting). Day 30 Discussion and recommendation during SAWP plenary (SAWP 2 meeting). Day 40 The CHMP final recommendation is adopted during the plenary meeting. Of note, requests related to Advanced therapy medicinal products (ATMPs) will also be circulated, after the SAWP, to the Committee for Advanced Therapies (CAT) for review and recommendation prior to finalisation and adoption by CHMP. The outcome, including the reasons that led to the CHMP's decision, will be sent by EMA to the applicant. An appeal mechanism is not foreseen.	EMA charges €63,000–84,000 for initial request and €32,000–42,000 for follow-up to the initial request. Fee reduction for small and medium-sized enterprises and for orphan drugs. Participating HTA agencies may also charge a fee.	Can include: • Scientific advice on key decision points/issues for the preparation of MAA with the potential to involve multiple stakeholders (e.g., HTA bodies, patients), when relevant. • Early appointment of CHMP/CAT Rapporteur (in line with current process, objective criteria, and methodology) • An initial kick-off meeting with multidisciplinary participation from the European network.

(Continued)

TABLE 5.1 (*Continued*)
Types of Early HTA Advice Offered by EMA and HTA Bodies

Type of Advice (Organization)	Prerequisites	Time Frame (Days)	Procedure Details Including Timelines (and Required Documents)	Fees	Specificities of the Process (Choice of HTA Bodies, Nonbinding, Key Focus of Questions, etc.)
Adaptive pathway (EMA and chosen national HTA bodies)	Treatments in areas of high medical need where it is difficult to collect data via traditional routes and where large clinical trials would unnecessarily expose patients who are unlikely to benefit from the medicine. In other cases EMA scientific advice should be pursued. • Conventional development pathway must not be decided • There must be iterative aspects of the development (conditional market authorization or expansion) • There must be a need to discuss development with HTA bodies • Real-world data must be considered for regulatory purpose	~50 minimum	• Timelines are flexible as this is a pilot project. • Applicants submit with EMA a proposal containing the main elements of the proposal: iteration, real-world data and HTA/patient interaction. Following EMA's reply, they submit a final Powerpoint presentation. • EMA has 14 days to set a date of a teleconference or a face-to-face meeting, which should happen not earlier than 28 days from the reception of the final presentation. • The teleconference can last 1.5–2.5h • One week after the teleconference the company should send minutes for record keeping. Minutes are not commented by EMA or HTA bodies.	n/a	• The teleconference cannot be considered a formal advice: there is no in-depth discussion of scientific aspects, that is, within the remit of a formal scientific advice procedure • Applicant can choose the HTA bodies that will participate • Confidential
National HTA advice (A national HTA body)	Country-specific procedures are described in the text in Section 5.2.				

In order to prepare for the HAS early meeting,[10] manufacturers should consider three conditions for eligibility as follows:

- Ongoing clinical development: results available for phase II studies and phase III study plan not yet initiated
- New therapeutic strategy (e.g., new mode of action)
- Unmet medical need

Remaining aspects of the application are described below.

5.2.1.1   Questions to Focus on

Manufacturers should focus on questions related to the development of the drug (i.e., comparator choice, modalities of administration, endpoints), but can also inquire about the therapeutic strategy of the disease in France, endpoint validity, quality of life assessment modalities, and so on.

For questions related to pharmaco-economic studies, manufacturers should focus on methodology choices such as types of analyses, included/excluded comparators, modeling perspective, population, time horizon, but also model choice, type of costs, and so on.

5.2.1.2   Process

Manufacturers willing to engage with this agency should follow the steps below:

- The company sends to HAS evidence to justify that the product meets the early advice eligibility criteria
- If the criteria are valid, the company receives meeting date suggestions from HAS
- First draft of the briefing package is then sent to HAS. A briefing book should include early stage data and methodology of planned phase III trials (if applicable, also pharmaco-economic study objectives and design)
- At this stage HAS may request questions or additional data
- Final briefing package should be sent at least 2 weeks before the planned meeting date

As an outcome of the meeting, the company should send to HAS the minutes of the meeting within 1 month. Minutes should include a summary of the context, background information on the disease, and target population, justification for seeking advice, company's question/position, HAS's answers to questions, comments, and conclusions. Finally, the minutes are validated by HAS.

Timelines of the early HTA advice with the French HTA agency HAS.

Steps of the Procedure	Timelines
• The company sends to HAS evidence to justify that the product meets the early advice eligibility criteria	–
• If the criteria are valid, the company receives meeting date suggestions from HAS	–
• First draft of the briefing package is then sent to HAS	–

(*Continued*)

Steps of the Procedure	Timelines
• HAS may request questions or additional data	–
• Final briefing package is sent by company to HAS	2 weeks before planned meeting date
• The company sends to HAS the minutes of the meeting	1 month after the meeting
• HAS validates the minutes of the meeting	–

5.2.1.3   Content of the Dossier

The Dossier should have the following structure:

- Background information on the disease to be treated including current management
- Background information on the product including product positioning and potential assessment from other agencies
- Efficacy and safety data from phase I and phase II clinical trials with levels of evidence
- Details of phase III clinical trial plan
- Protocol synopsis (at minimum)
- If applicable, the description of the pharmaco-economic study with expected effects in terms of health outcomes and costs
- If the economic assessment relies on an existing model, to provide publications presenting the economic model
- Questions and the company's position
- Questions related to study design, comparator, and endpoint choice (at minimum)

5.2.2   THE UNITED KINGDOM

NICE offers prospective and confidential early scientific advice.[11] Advice can be sought at any time. However, a useful time for requesting scientific advice could be during phase II studies before the planning of phase III studies. Only a limited number of advice slots are available.

Fees for the procedure vary, depending on the number and complexity of questions asked in the company briefing book (around £49,000 [+VAT] maximum).

NICE also offers a possible joint scientific advice meeting with the UK Medicines and Healthcare Products Regulatory Agency (MHRA). However, MHRA and NICE produce separate advice documents.

Manufacturers also have an option to request an advisory input from the clinical practice research datalink (CPRD), in order to gain knowledge of what real world data might be available either observationally or interventionally.

Remaining aspects of the application are described below.

5.2.2.1   Questions to Focus on

Manufacturers should focus on development strategies rather than preevaluation of data to support a submission to NICE. They can also inquire about the interpretation of NICE technology appraisal methods guidance and its relevance for the product's evidence development plans, research design considerations or preferences to support each proposed indication, economic evaluation design considerations or preferences, and about considerations and insights from existing models.

5.2.2.2 Process

Manufacturers willing to engage with this agency should follow the following steps below:

- The company requests a slot and should contact NICE at least 25 weeks before they wish to receive the final advice report
- NICE will send a copy of the standard contract to be agreed and signed by both parties at least 1 month before the briefing book submission
- Company sends the briefing book to NICE
- NICE confirms project size and total cost of the project
- Clarification questions are sent from NICE within 7 weeks of receiving the briefing book
- Company responds to NICE clarification questions within 2 weeks
- Face-to-face meeting with the company and NICE takes place approximately 11 weeks after the briefing book submission (at NICE premises) and will last 3 hours

As an outcome of the meeting, NICE sends a written advice report to the company approximately 7 weeks afterwards for a medium project or 9 weeks for a large project. Any possible clarification questions requested by NICE should be answered by the company within 15 working days. Finally, NICE should answer further company questions within 20 working days.

Timelines of the early HTA advice with the UK HTA agency NICE.

5.2.2.3 Content of the Dossier

The Dossier should have the following structure:

- Background information
- Disease and unmet needs
- Proposed product indication
- Treatment guidelines or recommendations
- Value proposition(s) for the product
- Data currently available on the product
  - A Brief description of the mode of action and pharmacological class
  - Proposed dosing regimen and route of administration
  - Data from completed clinical studies
- Proposed evidence plans for the product
  - Clinical trial designs, study population, comparator, endpoints, and the duration of observation
  - If available, a plan for the economic evaluation can be presented
- Questions and company's position
- Key references

Other key points to be considered when drafting the dossier are that any relevant information should be labeled as *commercial-in-confidence* (not as *confidential*), should not exceed 50 pages, including appendices, and should not include preclinical data.

5.2.2.4 Light Scientific Advice

Another type of advice offered by NICE is the so called *Light scientific advice*[12] that is designed for small to medium enterprises. The advice is a concise and quicker version of the standard scientific advice. Its key features are as follows:

- Provides answers to each of the company's key questions
- 12 week process from project start (additional 3 weeks for optional clarifications)

Steps of the Procedure	Timelines
• The company requests a meeting slot at NICE	At least 25 weeks before the company wishes to receive the final advice report
• NICE sends a copy of the standard contract to be agreed and signed by both parties	At least 1 month before the briefing book submission
• Company sends the briefing book to NICE	–
• NICE confirms project size and total cost of the project	7 weeks of reception of the briefing book by NICE
• Clarification questions are sent by NICE to the company	
• Company responds to NICE clarification questions	Within 2 weeks of receiving by the company NICE's briefing book comments
• Face-to-face meeting with company and NICE	11 weeks after the briefing book submission
• NICE sends a written advice report to the company	7 weeks for a medium project or 9 weeks for a large project
• Possible clarification questions requested by NICE are answered by the company	Within 15 working days
• NICE answers further company questions	Within 20 working days

Other requirements and documents needed are similar as those for the full scientific advice from NICE described above.

5.2.3 GERMANY

G-BA consultation is possible as per the SGB (German Social Code) V, section 35a, paragraph 7[13] in the following scope:

- On documents and studies to be submitted
- On the appropriate comparator

Additionally, Federal Institute for Drugs and Medical Devices (*Bundesinstitut für Arzneimittel und Medizinprodukte* [BfArM]) or the Paul Ehrlich Institute is involved if the product is evaluated before starting phase III studies.

Companies have to submit their request using a specific form, along with a cover letter.[14] Timelines for the procedure range from 4 to 5 months. Fees are between €5,000 and €10,000.

5.3 STRATEGIC CONSIDERATIONS

5.3.1 MULTIDISCIPLINARY APPROACH

Early HTA advice procedures require collaboration of multiple departments in the company, for example, preclinical development, regulatory, market access, internal representatives of the countries where participating agencies are located, medical affairs, health economics and outcomes research. They need to achieve clear strategic positioning, internal alignment, and maintain a proactive engagement to maximize the outcome of the advice.

5.3.2 BRIEFING BOOK IS THE CORNERSTONE OF EARLY HTA ADVICE

The briefing book is a document required in all types of early HTA advice. It gives a unique opportunity for the company to bring authorities' awareness to the disease and its related burden. It is crucial in order to develop a comprehensive background on the drug and its development plan. It is used to support discussions in advice meetings. Multidisciplinary approach will be required to draft the book.

The briefing book needs to incorporate accurate and specific questions (but not too narrow) so that detailed and clear answers can be obtained from the HTA bodies or regulators. It is also important to identify sensitive questions and review all questions or company's positions by external experts.

5.3.3 HOW TO CHOOSE THE RIGHT OPTION FOR EARLY HTA ADVICE

The company needs to identify the appropriate timing to seek advice. Broadly, this can be done in the following phases of product development:

- Very early in the drug development (nonclinical or proof of concept)

At this stage, the company may seek clarifications or adjustments of general clinical trial design but with limited patient data. The company is likely to obtain a general response with a less specific advice.

- Later in the drug development (prior to phase III)

Here, the company can obtain more precise responses regarding clinical trial design and pharmaco-economic questions. When phase III plans have been finalized, advice can still help to adjust design or statistical analysis plan of phase IIIb or IV studies.

The decision of when to consider the early HTA advice will depend on the following four main factors, assuming the product fulfils the criteria for the advice requested:

1. The development type envisaged: (i) traditional linear from phase I to III, (ii) medicine adaptive pathway to patients (MAPP), or (iii) a combined phase II or III development
2. The disease landscape
3. The product profile
4. The objective of the advice

Finally the risk related to involving in early HTA advice also depends on the type of advice chosen: (i) parallel, (ii) multi-HTA, (iii) national and on the time of advice in development such as (a) end of phase IIa, (b) end of phase IIB, (c) end of phase III, and (d) presubmission.

Although all those risks are interrelated, the risk analysis can be done in two steps: first, the time of advice, and second the type of advice. It is an artificial separation that allows us to conduct a stepwise analysis and therefore simplify the process. In the second stage when deciding on the type of advice, this may obviously have backward consequences on the time for advice.

This is a multidimensional decision process, where all dimensions are interrelated and impact each other. Except for the type of development chosen (traditional, MAPP, or combined phase II or III development) for which the options are independent and mutually exclusive, all others may be linked and coexist. Therefore, it is important to ensure that they are not double counted.

Overall, the advice should be sought early enough to ensure that the company can appreciate the advice from HTA bodies and integrate them in all phases of the development. If the advice is performed too early, population(s) and indication(s) may be dramatically affected by the requests of the HTA agencies and end of phase IIb should be a reasonable time to request for an advice. If the company has actual questions or misalignment on the population and/or indication, the end of phase IIa is recommended to seek for an advice.

End of phase III appears too late for an early HTA advice as most questions are already addressed in the phase III study. HTA bodies may feel bounded and unable to impact development decision because a large volume of evidence is already gathered and the room for influencing the product positioning in the development is minimal. However, early HTA advice at the end of phase III may support designing the phase IIIb study program.

In case of combined phase II or phase III, and in case of MAPP there is no option for end of phase IIb study HTA advice; therefore the end of the proof of concept study is the only reasonable early option for a fruitful dialogue that may inform the company's decision process and optimize the product's value through the development phase.

5.3.4 TYPES OF RISK FOR THE COMPANY WHEN CONSIDERING EARLY HTA ADVICE

When a company involves in an early HTA advice there are two potential risks involved: (i) target population related risks—the risk of the HTA and/or regulatory authorities asking to adjust the targeted population and eventually also the indication, as a result of the procedure and (ii) development plan related risk—the risk that the authorities request a major revision of the phase III trial or the overall development plan, as a result of the procedure. Even though the advice is nonbinding, the company will likely need to comply with the requests, in order to achieve unrestricted HTA recommendation once the drug is approved for use. Since such compliance can result in, for example, increased development costs and time, these outcomes should be considered as risks by the company. These risks are further discussed below.

5.3.4.1 Target Population Related Risk

HTA agencies may request to narrow the population indication because at the time of a future HTA review the evidence will be primarily gathered from the population they consider for reimbursement. This will put them in a more comfortable position to appreciate the product's value versus the appropriate comparator. Also, by avoiding situations where they need to restrict the population, they protect themselves from negative media coverage of the decision and from discontent of the patient groups and prescribers.

HTA agencies may consider restricting the product to more severe patient groups, to second or third line, to add on therapy, to patients with poor prognosis, or those at risk of safety issues that the product may avoid compared to its comparator. HTA agencies may also be willing to enlarge the population to prevent niching a product in a small group with high unmet need, thus leading to a high differential value. If the product is developed in a wider population, the added benefit may be more diluted although still present.

The risks increase when no development guidelines exist or when they are too vague, not specific enough, and when there are no historical products that may have paved the way and rendered a specific requirement for a new product.

Such risks are not constant over the product development stages. However, the more advanced the development is, the risk is less, but also the benefit from HTA consultation can be smaller. Delaying the consultation reduces the risk, but also reduces the benefit. This is a trade-off to consider.

The risk is very high at the end of the phase IIa study because it is affecting the target population. As the phase IIb has not yet been performed, there is still room to revisit the target population and the indication. At that stage, all decisions are still possible. So HTA agencies may be more assertive at that stage to impact the population or indication.

After the end of the phase IIb study, the risks diminish dramatically because in order to achieve a marketing authorization approval, usually two studies are needed in the same population. The phase III is expected to replicate the result of the phase IIb in a reasonably similar population. Even if some adjustments are possible they should not be dramatic. If one significantly changes the population, one may be requested by regulators to redo their phase IIb study. Therefore the risk of a dramatic change of the population targeted for the indication is very small.

At that stage, HTA agencies can push to enlarge the included population in the phase III by introducing wider inclusion criteria and less strict exclusion criteria. For example, include older patients, more severe patients, accept more comorbidity, or accept more comedications in order to be closer to the population that will be treated in real life.

At the end of phase III, the HTA consultation is unlikely to be a multiHTA advice as this scheme assumes that HTA agencies can influence the phase III trial design. However, it may be possible in theory to apply for a parallel advice.

The expected impact on the population and/or indication at the time of presubmission is null.

5.3.4.2 Development Plan Related Risk

HTA agency consultation may lead to dramatic changes in the development plan. This may affect the phase III study only, but it may also affect the overall development plan, leading to additional unexpected studies that otherwsie would not be requested by the regulatory agency.

HTA may request a broader phase III study, but also a broader development plan. This risk is similar at the end of proof-of-concept and at the end of phase IIb, although smaller at the end of phase IIb.

Such adjustments may delay registration and/or reimbursement.

Also when undergoing a scientific advice, there is some risk that the HTA agencies require substantial changes either in the phase III study or in the development plan as a whole. Of course this risk diminishes as the development is more advanced.

The HTA body may request substantial changes in the development plan and the phase IIb and phase III design. Among the HTA request that may influence the phase IIb and/or phase III trial, we can consider the following:

Introduce an active comparator, introduce a second comparator, introduce an additional arm(s) with different treatment regimen(s), enlarge or narrow the inclusion or exclusion criteria, additional end point, change the calculation method of the primary or secondary end points, lengthen the duration of the study, lengthen the post dosing follow-up, request a noninclusion patients registry with a baseline characteristic and so on.

Such risks are difficult to appreciate, and their appreciation is based on the company expertise and experience. It has to be confronted with external experts' opinion.

Such advise could be an opportunity to design a phase IIIb study that would specifically address the HTA agencies' expectations.

One possible reason to seek advice at the end of phase III is, when the phase III revealed an unexpected benefit. Then, the company may need advice for optimizing the value of this benefit. Three types of advises are possible:

- How to design an acceptable study for HTA to optimize the appreciation of the benefit
- Which observational study will allow to quantify the magnitude and the scope of the problem addressed through this benefit
- How this benefit could be translated in a cost-effectiveness advantage

At that stage, HTA agencies requirements may impact the reimbursement but not the registration. It can be seen as a way of show a goodwill with HTA agencies although not affecting the registration process.

Finally, seeking an advice at the time of presubmission to HTA agencies can help to raise awareness about the condition, understand the resistance of HTA agencies to the way the evidence is presented and overcome the resistance by addressing them, or adjusting the evidence package.

It may also be an opportunity to collect information through database analysis, chart abstraction, or cross-sectional design to bring evidence that will address the resistance while the dossier is under review.

5.4 CONCLUSIONS

The main goal of the early HTA advice is to achieve consensus between HTA bodies and the EMA (when relevant) on the global drug clinical development plan. Simultaneous feedback from HTA bodies and regulators can help companies to identify key areas of consensus and divergence between these different stakeholders.

REFERENCES

1. EMA website. Health-technology-assessment bodies. Available from: http://www.ema.europa.eu/ema/index.jsp?curl=pages/partners_and_networks/general/general_content_000476.jsp&mid=WC0b01ac0580236a57 (accessed September 29, 2016).
2. EMA Presentation. HTAs and EMA working together: 23 parallel scientific advice procedures later-wwhat have we learned? DIA 26th Annual EuroMeeting, Vienna 2014. Presented by Jan Regnstrom. Available from: http://www.epaccontrol.com/common/sitemedia/PrePost/PostPDFs/1035588.pdf (accessed September 29, 2016).
3. HAS website. Available from: http://www.has-sante.fr/portail/jcms/c_1700958/fr/seed-shaping-european-early-dialogues-for-health-technologies (accessed September 29, 2016).
4. HAS SEED Procedure. Available from: http://www.has-sante.fr/portail/upload/docs/application/pdf/2014-03/procedure_for_seed.pdf (accessed September 29, 2016).
5. EUnetHTA website-Joint Action WP5—Relative Effectiveness Assessment of Pharmaceuticals. Available from http://www.eunethta.eu/activities/JA-WP5/ja-wp5-relative-effectiveness-assessment-pharmaceuticals (accessed September 29, 2016).
6. EUnetHTA website- The final version of HTA Core Model® and the Methodological Guidelines for Rapid REA of Pharmaceuticals. Available from: http://www.eunethta.eu/news/final-version-hta-core-model-and-methodological-guidelines-rapid-rea-pharmaceuticals-now-availa (accessed September 29, 2016).
7. EMA website-Adaptive pathways. Available from: http://www.ema.europa.eu/ema/index.jsp?curl=pages/regulation/general/general_content_000601.jsp (accessed September 29, 2016).
8. EMA website-Priority medicines (PRIME) scheme. Available from: http://www.ema.europa.eu/ema/index.jsp?curl=pages/regulation/general/general_content_000660.jsp&mid. (accessed September 29, 2016).
9. EMA website- Priority Medicines Scheme. Available from: http://www.ema.europa.eu/ema/index.jsp?curl=pages/regulation/general/general_content_000660.jsp&mid=WC0b01ac058096f643 (accessed September 29, 2016).
10. HAS website. Available from: http://www.has-sante.fr/portail/jcms/c_1625763/fr/deposer-une-demande-de-rencontre-precoce (accessed September 29, 2016).
11. NICE website-Scientific advice-Available from: http://www.nice.org.uk/about/What-we-do/Scientific-advice (accessed September 29, 2016).
12. NICE website-Light Scientific Advice-Available from: https://www.nice.org.uk/about/what-we-do/scientific-advice/light-scientific-advice (accessed September 29, 2016).
13. G-BA website-The benefit assessment of pharmaceuticals in accordance with the German Social Code, Book Five (SGB V), section 35a-Available from: http://www.english.g-ba.de/benefit-assessment/information/#9 (accessed September 29, 2016).
14. Anlage I zum 5. Kapitel—Anforderungsformular für eine Beratung.

6 Overview of Market Access Agreements

6.1 BACKGROUND

Market access (MA) for drugs defines the rules that apply to allow a pharmaceutical company to enter a specific market. Those rules could be classified as generic rules that apply for all products and specific rules that aim to optimize a patient's access to a specific drug. In this chapter, we discuss the latter case, and we call such arrangements between the manufacturers and the payers market access agreements (MAAs).

Even though MAAs have been increasingly implemented over the past few years, confusion remains about the definition, taxonomy, and best practice for these schemes.

Historically, there have been nonwritten agreements between manufacturers and payers on how to share the risk of the drug development and commercialization. Manufacturers would bear the development cost and the risk of development failure. Once the product reached the market and a price was set between payers and manufacturers, the payers endorsed the postapproval risk. Currently, MAAs have changed this landscape as these schemes aim at minimizing the postapproval uncertainty for the payer and shift it toward the manufacturer. This was mainly driven by increasing prices of novel pharmaceuticals over time. Payers have become concerned about the effectiveness uncertainty associated with high-price products.

It is interesting to note that MAAs have been increasing in the recent years. Between 2010 and 2011, their number nearly doubled the total number of MAAs found in 2009. Furthermore, it is a growing phenomenon across countries, such as Italy and Australia, and a significant number of new countries such as New Zealand, Belgium, Poland, and Hungary[1] are beginning to use MAAs.

In this chapter, we will shed the light on these agreements, their different types, why and how they are put in place in different countries, the outcomes, and finally the possible perspectives for MAAs.

6.2 RATIONALE BEHIND MAAs

- It is difficult for the payers on the pharmaceutical market, under budget constraints, to manage the chasm between the expected benefit and the actual benefit of a new product.
- The clinical development plans of new molecules are not sufficient to assess the actual benefit before the launch of a new drug.
- MAAs are sometimes disguised discounts conceded by the pharmaceutical manufacturers to account for the effectiveness uncertainty inherent to the launch of a new product.

It was the budget constraint, as well as experiences from financing expensive drugs, that pushed payers to become more and more critical when assessing the value of new, costly drugs.

On many occasions, the expected benefits of innovative drugs were not met once the drug was on the market, which emphasized the important gap between the price and the drug's true benefit.

This gap between "expected benefit" and "actual benefit" was difficult to manage for payers, which led them to demand higher levels of evidence on the value of new drugs.

Traditionally, clinical development plans of new molecules aim to demonstrate the safety and clinical benefit of the product. However, they rarely allow anticipating the added value from a payer's perspective.

Most often, even if this added value is likely, it is not always evidenced before the launch, and uncertainty on the added benefit remains.

It is in this context that numerous agreements between payers and manufacturers were made.

It is interesting to note that most of these MAAs are in fact disguised price discounts to which both the payers and the manufacturers agree. As such, it appears reasonable that, as long as the uncertainty on the drug's value exists, the price accounts for this uncertainty or is conditional until the benefit is evidenced. Hence, we can already consider two types of agreements: one in which the price includes the uncertainty, and the other in which the price is conditional until the benefit is evidenced. There can also exist agreements that are a mix of the above two kinds.

6.3 DIFFERENT DEFINITIONS AND TAXONOMIES OF MAAs

6.3.1 DIFFERENT DEFINITIONS

- The definition of MAAs is controversial. They are usually referred to as risk-sharing agreements and are, often mistakenly, equated with cost sharing and payment for performance (P4P).
- In reality, MAAs are a mix of financial agreements and P4P.
- First De Pouvourville and later Towse and Garrison proposed definitions of MAAs with a focus on risk sharing. According to them, these agreements allow for the two parties of a transaction to share the risk that is inherent to the launch of a new product. They called attention to the distinction between laboratory efficacy and real-life effectiveness with respect to the appraisal of performance.

- Adamski et al. proposed a wider definition of these agreements with a focus on their cost containment capacity and a broader approach of the notion of risk.
- The UK health authorities presented a Patient Access Scheme (PAS) formula that occasionally resulted in a restricted access in defiance to its designation.
- Broadly, all MAAs are the result of a settlement between health-care payers and the industry on the drug's price and reimbursement status, health technology assessment (HTA) recommendation, and/or formulary listing.

To date, there has been a confusion surrounding the exact definition of MAAs, and there is no commonly agreed definition for MAAs. In fact, a number of definitions can be found in the literature as well as different taxonomies.

MAAs are often referred to as risk-sharing agreements, despite the lack of risk sharing in most of them. Consultants call them innovative contracting, whereas in the United Kingdom, a new terminology was adopted by the Department of Health (DoH) that is the Patient Access Scheme (PAS).

Risk sharing, cost sharing, and P4P are often put in the same basket and called risk-sharing agreements, whereas in reality, there are some structural differences. Some often quoted definitions are proposed as follows:

- *Risk-sharing agreement*: "A contract between two parties, who agree to engage in a transaction in which there are uncertainties, regardless of its final value. Nevertheless, one party, the company, has sufficient confidence in its claims of either effectiveness or efficiency that it is ready to accept a reward or a penalty depending on the observed performance of its product."[2]
- *Cost-sharing agreement*: "Type of a Commercial Agreement between health care payer and the drug manufacturer implemented to reduce pharmaceutical

expenditure. The cost of the drug (according to its list price) is "shared" between the two parties either for all patients on treatment, or for patients, who did not respond to the treatment. In the later case it is a part of a Payment by performance agreement."[3]
- *P4P*: "A performance-based agreement can be defined as one "between a payer and a pharmaceutical, device or diagnostic manufacturer, where the price level and/or revenue received is related to the future performance of the product in either a research or a real-world environment."[4]

In practice, many MAAs are a mix of a financially based agreement and P4P.

Examples of these mixed types of agreements in Italy is given in Table 6.1.

Chronologically, De Pouvourville was first to define pharmaceutical risk sharing in the literature as "a contract between two parties, who agree to engage in a transaction in which there are uncertainties regardless concerning its final value. Nevertheless, one party, the company, has sufficient confidence in its claims of either effectiveness or efficiency that it is ready to accept a reward or a penalty depending on the observed performance of its product."[2] For example, the reward could be a higher price, or an extension of a license, and a penalty could be a lower price, with or without the reimbursement of excess profit in case the claims are not justified. Under such an arrangement, one can talk of risk sharing because parties, the company, and the payer support the financial consequences of reducing uncertainty. Such an arrangement necessarily includes the design and the cost of the observational procedure to assess the performance of the product in real life.[2] It is interesting that the author restricts the scope of his definition to risk sharing that represents a very small proportion of agreements. He focuses explicitly on risk management for the payer side. Towse and Garrison also call such agreements "risk sharing" and focus on the uncertainty of product performance—namely, these are "agreements between a payer and a pharmaceutical company where the price level and/or revenue received are related to the future performance of the

TABLE 6.1
Examples of MAAs in Italy for Oncological Drugs

Cost sharing		Sunitinib (Sutent), Erlotinib (Tarceva)
Discount of 50% on the drug price for the first cycles of therapy	Description of the scheme	50% discount on sunitinib or erlotinib for the NHS for the first 3 months (two cycles) of treatment
Risk sharing		**Dasatanib (Sprycel)**
A patient is treated at full costs until follow-up; if the patient shows disease progression, then the manufacturer has to pay back 50% of the treatment cost	Description of the scheme	50% price reduction for patients with disease progression after the first month/cycle of treatment
Payment by performance		**Nilotinib (Tasigna)**
A patient is treated at full cost until follow-up; if the patient shows disease progression, then the manufacturer has to pay back the full cost	Description of the scheme	Full price for the first month of treatment; then 100% payback for nonresponding patients

Source: C. Jommi, *Central and Regional Policies Affecting Drugs Market Access in Italy*, Bocconi University, Milan, Italy, 2010.

product in either a research or real-world environment."[4] They clearly differentiate the efficacy and effectiveness by referring to research and real-world evidences.

Adamski et al.[5] proposed a wider perspective and called those agreements risk-sharing arrangements. They proposed the following definition: "agreements concluded by payers and pharmaceutical companies to diminish the impact on the payer's budget of new and existing medicines brought about by either the uncertainty of the value of the medicine and/or the need to work within finite budgets." In practice, the agreement lies in setting the scope and realizing the mutual obligations among both payers and pharmaceutical companies depending on the occurrence of an agreed "risk." The risk varies and can include pharmaceutical expenditure higher than agreed thresholds or a health gain from a new product lower than that shown in clinical trials. Adamski et al. explicitly focused on the cost containment role of such agreements, and the risk lies more on the budget impact side than on the intervention performance side. It is noticeable that he himself and his coauthors are payers, and focus on the budget impact, whereas the authors, who focus on uncertainty of the performance, are academics.

Thus, there is no consensus regarding the definition of MAAs. They may be defined as an outcome of a compromise between health-care payers and the industry on the drug's price and reimbursement status, HTA recommendation, and/or formulary listing. Hence, this definition may overarch all the types of agreements introduced so far.

In the De Pouvourville's definition, agreements, that only aim at cost containment are not included. In such agreements, there is no uncertainty about health outcomes. These financial agreements are also often called risk-sharing agreements, although most do not imply sharing a risk. In fact, in most identified agreements, the risk is shifted solely on the manufacturer, rather than being shared between both parties.

PAS is another MAA term that was proposed by the UK's DoH. PASs are defined as "innovative pricing agreements designed to improve cost-effectiveness and facilitate patient access to specific drugs or other technologies." However, some schemes have led to restricted access rather than optimizing the patients' access.

6.3.2 A Possible Definition

- In this book, MAA is defined as "an agreement between two or more parties, who agree on the terms and conditions under which a product will get access to the market." This definition covers both the financial and outcome-based types of agreements.

In this book, we use the term "market access agreement" (MAA), as it covers all types of agreements.

MAA is defined as "an agreement between two or more parties, who agree on the terms and conditions under which a product will get access to the market."

It can be grouped into two types of agreements: financial agreements and outcome-based agreements. The latter ones include both P4P and coverage with evidence development (CED) agreements, which are all presented in Section 6.3.3.

6.3.3 Different Taxonomies

- There are as many classifications as there are definitions of MAAs. They cover a wide array of criteria, such as financial and performance-based/outcome-based models and population/patient-level schemes.
- Carlson et al. proposed a thorough, inventory-based taxonomy. However, it remains complex for daily practice and classification of agreements.

In addition to the various terminologies describing these schemes, experts have proposed different classifications.

Adamski et al. recommended that such agreements may be classified into two categories: financial and performance-based/outcome-based models. Other classifications distinguish between population- versus patient-level schemes, the latter characterizing a situation in which the value for money is guaranteed without the need for subsequent review of the reimbursement decision.[5]

Towse and Garrison also made the distinction between outcome-based and non-outcome-based schemes in their taxonomy, and also between those that specified how evidence would be translated into revisions to price, revenues, and/or use, and those that instead specify an evidence review point where renegotiation would occur.[4]

Carlson et al. based their taxonomy on an inventory of published schemes categorized in terms of timing, execution, and health outcomes, and made a clear distinction between schemes that are health outcomes based and those that are not, setting aside the research component.[6]

This taxonomy allows identifying the various subtypes of agreements. It is useful to review all potential agreements that might apply for a given intervention to prepare for negotiation with payers. However, it remains complex for daily practice and classification of agreements (Figure 6.1).

6.3.4 Simplified Taxonomy

- MAAs can be classified into two distinct groups: financial agreements concerned with the financial aspect of the transaction and outcome-based agreements which amount to a covenant to deliver an expected added benefit.
- Two major subgroups are population-level and individual (patient)-level MAAs.

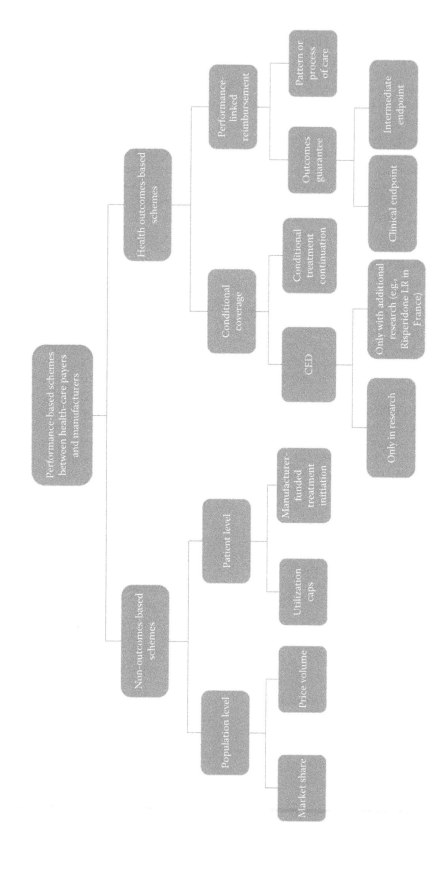

FIGURE 6.1 Taxonomy of performance-based health outcomes reimbursement schemes. (Adapted from J.J. Carlson et al., *Health Policy*, 96, 179–190, 2010.)

TABLE 6.2

Taxonomy of MAAs

	Population Level	Patient Level
Financial based	Price/volume agreement	• Capping • Manufacturer-funded treatment initiation
Outcome based	CED	• P4P reimbursement

Source: S. Jaroslawski and M. Toumi, Market access agreements for pharmaceuticals in Europe: Diversity of approaches and underlying concepts, *BMC Health Serv Res.*, 11, 259, 2011.

MAAs can be grouped into two types: financial agreements and outcome-based agreements:

- "Financial agreements are commercial agreements between two or more parties entering into a deal for goods acquisition. For example, a price volume agreement, discount or rebate, cap price, or cost sharing."[7]
- "Outcome-based agreements are part of an insurance or warranty facility: the payer agrees to a price under the insurance that the product will deliver a predefined outcome. This regroups two kinds of MAA: Payment for Performance (P4P) and Conditional coverage agreements."[7]

These two types of MAAs are also subdivided in two categories: MAAs at the population level and MAAs at the individual patient level, as shown in Table 6.2.

6.3.4.1 Commercial Agreement

- Financial agreements are concealed price discounts and do not involve any kind of risk sharing.

Financial agreements are commercial agreements (CAs) between two or more parties entering into a deal for goods acquisition. The aim of CAs is cost containment, and it does not involve any risk sharing.

In order to maintain high list prices (which are used internationally by payers as reference when setting a price for a new drug), pharmaceutical companies have been entering into more and more complex agreements with payers, which conceal the actual price at which the drug is purchased.

This can be achieved by implementing price–volume mechanisms and market caps. Most of these agreements are more or less complex price discounts.

6.3.4.2 Payment for Performance

- In P4P agreements, the payment hinges on the outcome observed in individual patients.
- Conditional treatment continuation is a type of P4P in which continuation of coverage for individual patients is conditioned upon meeting short-term treatment goals.

- Prospective P4P (PP4P) is another type of P4P in which payment is made on a periodic basis for as long as the patient benefits from the product.
- PP4P tapers the uncertainty level and the budget impact and can accelerate the implementation of new health technologies that otherwise would not be financed by a payer due to high cost.
- PP4P however ascribes the whole risk to the payers. It also provides little information on the added value of the product at the time of launch. Furthermore, it requires time to generate evidence, as well as thorough monitoring and substantial administrative costs.

When the payer agrees to a price under the insurance that the product will deliver a predefined outcome, it is an outcome-based agreement. In a P4P agreement, payment is decided for individual cases, which allows minimizing the possibility of paying for nonresponders and maximizing the likelihood that the manufacturer receives a price reflecting the drug's true value.

The analysis is done on an individual basis and is potentially infinite in duration. P4P does not seek to gather new or comparative evidence but rather to justify the reimbursement of a drug in individual patients.

A product is only eligible for PP4P if it is proven to be innovative, with a high health outcome potential, expensive, and ideally for short-term use. In this situation, capital cap is also likely to be implemented.

6.3.4.2.1 Conditional Treatment Continuation

Among P4P agreements, we can also distinguish a type of P4P in which continuation of coverage for individual patients is conditioned upon meeting short-term treatment goals (e.g., tumor response or lower cholesterol level). This form of P4P was termed conditional treatment continuation by Carlson et al.[6]

6.3.4.2.2 Prospective Payment for Performance

The authors are also aware of a different type of P4P agreement that they propose to call "prospective payment for performance." It has recently been implemented in oncology for two different products. The specific conditions of the schemes cannot be revealed as they are protected by a confidentiality agreement; however, it is interesting to describe the rationale and the pros and cons of such an agreement.

In this model, the insurance would pay on a periodic basis (e.g., quarterly) for as long as the patient benefits. It could be linked to progression free survival in oncology or until next flare or relapse in ulcerative colitis, and so on.

The advantages of this type of agreement would be the following:

- The reduction of decision uncertainty inherent to purchasing new technologies by replacing a single upfront payment with a payment stream with an equivalent expected value conditional upon delivery of predefined health gain

- The reduction of the initial and expected total budget impact of the technology through the progressive reimbursement process
- The prompt and easier implementation for innovative technologies into a healthcare system

The limits of such scheme are as follows:

- The uncertainty/risk is shifted on the payer, who covers a drug that is yet to prove its real-life effectiveness. The payer could only pay for proven effectiveness or cover new dug with additional evidence development; instead, here the payer will pay for it despite no specific evidence at the time of launch and will learn about the drug's added benefit only later.
- The lack of comparative study makes it difficult to assess if the given product is doing better than a reference product in real life. Patient's eligibility criteria should be very strict and documented, or some patients (e.g., those with a light degree of a condition) might mistakenly be treated with the new technology and the apparent added benefit would be confounded by the good basal patient status.
- Such scheme does not generate evidence over a limited period of time and will likely last as long as the drug is covered by patent protection and thus has high price or untill novel drugs are available.
- The potentially substantial administrative costs associated with monitoring the patients' response to treatment.

The conditions for eligibility of PP4P are as follows:

- It is a new treatment for pathology with no effective treatment, which demonstrated a good benefit with a high, but unproven therapeutic potential.
- Ideally, the treatment duration should be short, coupled to a long-lasting health benefit after the end of the treatment.
- The drug must belong to a range of highly priced drugs to compensate for the substantial administrative costs associated with monitoring over time.
- It is likely that cap payment will need to be implemented to avoid paying without time limits.

6.3.4.3 Coverage with Evidence Development

- CED is a type of agreement in which the payer grants conditional funding for a product, whereas evidence is being gathered through a study under the conditions consented by both parties.
- They vary in terms of visibility and give way to price negotiations as new data become available from the ongoing studies.

- They are by nature temporary agreements. They offer cost-effectiveness safeguards for the payer and grant the manufacturer a premium price for their product conditionally.
- In CEDs that involve an escrow agreement, the payer can claim a refund if the product fails to meet the expected level of benefit.
- They aim to minimize the uncertainty in regard to the study results as well as the real-life benefit of the drug.

CED is a conditional outcome-based agreement. It allows for conditional funding of new promising drugs, whereas more conclusive evidence is being gathered to address uncertainty regarding clinical or cost-effectiveness at a population level.[8]

CED agreements suppose the demonstration of a health benefit in a study to which both parties agreed. The threshold of the particular benefit may be predefined. In other cases, different commercialization prices are predefined, depending on the level of the benefit.

Finally, the money paid to the manufacturer during the testing period could be claimed as payback, in case the performance of the drug does not meet the predefined expectations. Anglo-Saxon legal experts call this type of schemes "escrow agreements." In an escrow agreement, the amount due by the payer to the manufacturer is put aside on a frozen bank account or placed with a notary. If, by the end of the study, the results confirm the drug's performance according to the scheme, then the frozen amount is given to the manufacturer. Otherwise, it is returned to the health-care payer.

The scheme visibility also varies from one CED to another. Even though CEDs have a limited duration, it is not always clear at the time of implementation how much the price revision will be, because the minimum threshold of clinical improvement and the relation between magnitude of improvement and price are not defined.

The aim of CED agreements is to minimize the possibility of financing non-cost-effective drugs and maximize the likelihood that the manufacturer receives a price reflecting the drug's true value.

CED agreements are temporary agreements that are terminated, when evidence from a cohort of patients is produced. Their temporary nature allows for developing new evidence on real-life health outcomes and identifying elements such as the duration of treatment, the appropriate dose, and the target population.

CED aims to reduce two types of uncertainties:

- How efficacy results in clinical trials translate in real-life benefit
- The clinical study design, for example, providing more conclusive evidence about the real value of the product versus a comparator rather than placebo (Table 6.3)

TABLE 6.3
Classification of MAAs

MAA Category	CA	P4P Agreement	CED
Contract type	• Traditional commercial contract	• Long-term risk-shifting agreement (outcomes guarantee) applied on a per-patient basis	• Provisional agreement (risk sharing or shifting) until new evidence develops from a cohort of patients
Underlying concept (payer's perspective)	• Reducing pharmaceutical expenditure	• Avoiding inefficient expenditure on treating patients who do not respond to a drug and who cannot be identified *ex ante* (by infinitely linking the payment to drug's performance)	• Avoiding inefficient expenditure until uncertainty about drug's effectiveness is reduced (by linking final reimbursement (and/or pricing) decision to drug's performance)
Types	• Price–volume agreement • Market cap • Flat price (per patient, regardless of the number of doses administered) • Cost sharing • Rebate • Discount	• P4P • Payback for nonperformance • Payment for management of events which the drug failed to prevent • Payment for management of side effects	• Temporary coverage on a condition that new evidence reduces uncertainty about • Real-life effectiveness • Higher efficacy in a subpopulation of patients • Actual daily dose • Long-term effect • Improved patient's adherence • Reduction of use of health-care resources • Reduction of use of other medication having serious side effects

Source: S. Jaroslawski and M. Toumi, Market access agreements for pharmaceuticals in Europe: Diversity of approaches and underlying concepts, *BMC Health Serv Res.*, 11, 259, 2011.

6.4 PAYERS' AND MANUFACTURERS' MOTIVATIONS TO IMPLEMENT MAAs

- Financial agreements and P4P agreements are budget control tools, whereas CED is a specific tool for managing uncertainty.

MAAs have emerged as an answer to two issues: restricted budget and uncertainty surrounding the real benefit of the new product.

In reality, some represent tools for budget constraint management and others for the management of payer's uncertainty about drugs' added benefit.

CED is a specific tool for managing uncertainty, whereas CAs and P4P agreements are used to control the budget and pharmaceutical expenditure.

6.4.1 THE INCREASINGLY COST-SENSITIVE ENVIRONMENT

- The development of MAAs is driven by the strains put on the payers' budgets.

The increasing number of new, expensive technologies that are supported by the state's budget has put considerable strain on European health-care systems, most of which are in deficit.

The growing cost pressures on health-care budgets result in cost containment measures of authorities, such as these agreements.

6.4.2 THE UNCERTAINTY RELATED TO DRUG'S PERFORMANCE

- The development of MAAs is also driven by the growth of the payers' requirements in terms of evidence of the value of new products.

In most developed countries, public authorities and payers are putting increased emphasis on the postmarketing monitoring of drugs. Pressures on health-care budgets certainly drive this change, but this emphasis extends farther than merely budget impact and pricing issues.

In contrast to the initial paradigm that prevailed in the 1990s, public authorities and payers are now questioning the effectiveness of new products: the clinical efficacy demonstrated through the regulatory approval is considered a necessary but not sufficient condition to verify the value of a new product.

In fact, clinical trials do not demonstrate effectiveness due to a number of biases (experimental setting, selection of patients in the trial, choice of the comparator, the intermediate endpoints, the duration of the trial, etc.)[2] to which payers and health authorities are becoming less tolerant.

6.4.3 THE "TRUST CRISIS"

- The payers are short of trust in the pharmaceutical industry, and MAAs allow them to tally the price with the real value and condition their decisions on the true performance of a drug.
- MAAs meet the payers' need to control spending and the manufacturers' need to achieve high list prices. However, although the drug efficacy data produced can benefit the payers, it may be detrimental to the manufacturer.

In the current pharmaceutical industry crisis, there is a need to restore trust, as the uncertainty related to drug's performance has become a critical issue for payers. Even though the risk often appears to be linked to pharmacovigilance (monitoring side effects), the uncertainty related to lower-than-expected performance in real life is frequent. In that case, MAAs can allow relating the price of the drug with its real added value. It can also allow the payers to reverse or review pricing and reimbursement decisions more easily.

In practice, while taking such decisions, the payers' main goal is to control expenditure and provide patients with access to new drugs. The manufacturers aim to keep high list prices

for drugs in order to optimize external (international) reference pricing (Chapter 7).

There are, however, a number of other reasons to implement MAAs on both sides, which are presented in Table 6.4.

Some other benefits that could motivate payers to implement MAAs are to collect valuable postmarketing data on the drug and its benefits over time with possible increased coincidental information, which could be useful for another product.[2]

Some risks may arise when applying MAAs:[2]

- Manufacturers may produce data that are useful to competitors, (i.e., second movers on the market in the therapeutic area).
- Manufacturers may fail to deliver evidence requested by payers as a part of CED and continue enjoying reimbursement at a premium price without negative consequences.

Any plans to establish the MAA should be preceded by a comprehensive and transparent analysis of stakeholders' motivations, and the MAA contract should precise what happens if either of the parties does not keep the terms of the agreement.

6.5 INTERNATIONAL COMPARISON OF MAA HEALTH POLICIES

6.5.1 MAAs ACROSS COUNTRIES

- MAAs are very common in Europe, and there is a large variety in implementation.

MAA types differ across countries and their implementation differs as well.

TABLE 6.4

Payers' and Manufacturers' Motivations to Implement MAAs

Payers' Motivations	Manufacturers' Motivations
• To provide patients with access while staying within budget	• To optimize international reference pricing
• To control expenditure	• To capture the product's value (ensure profitability)
• To improve ICER of expensive products (and stay within a fixed threshold)	• To reduce the cost of an additional clinical trial
• To prioritize certain interventions to patients who most need it	• To achieve access for patients and further explore drug's potential
• To align with a national/regional policy that prioritizes certain health outcomes	• To achieve competitive advantage (over cheaper or equally priced comparator drugs)
• Under pressure from patient associations, government, or media	• To capture value of nonclinical or real-life benefits of the drug
• To prevent media coverage of negative reimbursement decision	• To obtain comparative real-life effectiveness data versus a comparator
• To expand the list of innovative accessible interventions in the country (favorable publicity)	• To mitigate a failure to achieve reimbursement or HTA recommendation and achieve global coverage (for policy reasons)
• To reduce uncertainty about real-life effectiveness or nonclinical drug benefits (e.g., compliance)	• To improve company's image
	• "Opportunity to develop partnerships with payers"
	• For example, Roche has promoted a number of educational programs on oncology registry and MAA for clinicians and hospital pharmacists

- Most CED and P4P schemes were implemented outside the United States, between 1998 and 2009.[6]
- MAAs are very common in Europe and there is a large variability in schemes' implementation. These are presented in Table 6.5.
- The price–volume schemes are extensively implemented by the industry due to the fact that they can help them achieve relatively high volumes of sales. They are used by payers either as a leverage to control drug use in target populations or as a tool to control the budget impact.
- In practice, a mix of CAs and P4P is agreed upon and implemented.
- CAs are most common in France, Spain, and Italy.
  - Price–volume agreements in France do not apply on the first box that is sold, but rather in an incremental way, with prices varying when crossing a certain volume. In most countries, these agreements are implemented from the first box that is sold, meaning that the sales volume reached over a period of time will condition the unit price and apply to the total sales.

- In Italy, the use of MAAs has increased (Figure 6.2). They are mainly negotiated and decided at the national level with the Agenzia Italiana del Farmaco/ Italian Medicines Agency (AIFA), and primarily managed at the regional level. The lack of transparency and scientific grounds for MAA construction makes Italy an unpredictable market.
- CED schemes are usually implemented for costly drugs in Belgium, the Netherlands, Sweden, and Portugal. They are increasingly being used in France after a long period of reluctance.
  - Initial reimbursement or pricing decision is reviewed after 1.5–3 years.
- Sweden is the most active country to implement CED schemes. Swedish Dental and Pharmaceutical Benefits Agency (TLV), operating under the cost-effectiveness paradigm, does not reimburse drugs with uncertain or high incremental cost-effectiveness ratio (ICER).
  - For instance, levodopa/carbidopa intestinal gel went through a 5-year-long CED scheme aiming at reducing uncertainty of its ICER.

TABLE 6.5
Differences in MAAs across Some European Countries

MAAs	FR	UK	DE[a]	IT	ES	NL
CAs						
Market cap	√	√		√√	√	
Price volume	√√		√	√		√
Flat price	√	√	√		√	√
Discount	√	√	√	√	√	√
Rebate	√	√	√	√	√	√
Cost sharing		√	√	√		
P4P						
Payback for nonperformance		√	√	√	√	
P4P		√	√	√	√	
Payment for nonprevention management			√			
Payment for side effects management			√			
CED						
Real-life effectiveness	√	√	√			√
Subpopulation efficacy	√	√	√			√
Long-term effectiveness	√	√	√			√
Actual daily dose	√		√			√
Improved adherence	√		√			
Resource use reduction	√		√			
Comparative clinical trials	√	√				

Notes: FR-France; UK-United Kingdom; DE-Germany; IT-Italy; ES-Spain; NL-the Netherlands.

[a] Market characterised by fragmented private health insurance.

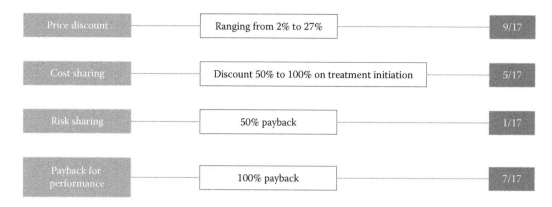

FIGURE 6.2 Typology of MAAs in Italy. (17 evaluated agreements.) (From C. Jommi, *Central and Regional Policies Affecting Drugs Market Access in Italy*, Bocconi University, Milan, Italy, 2010.)

- The product was given at launch a price and a reimbursement rate that were conditional on the company carrying out a study. However, the post-marketing study was conducted without the TLV's approval on the design, and the results were not relevant for TLV. For that reason, levodopa/carbidopa intestinal gel was not granted reimbursement at the convened conditional price. Afterward, a second study was conducted with the TLV that allowed to evidence the added value of levodopa/carbidopa intestinal gel and to confirm reimbursement at the initial conditional price.
- MAAs in Germany have the following features:
  - Primarily at the Länder (regional) level: contracting with the Krankenkasse (sickness funds)
  - Employ a mix of methods (price–volume, outcome-based, etc.)
  - Increasingly used by private insurers
- MAAs in the United Kingdom have the following features:
  - PASs: the largest and most formal program
  - It is run at the national level (National Institute for Health and Care Excellence [NICE]/National Health Service [NHS])
  - The focus is to drive cost-effectiveness by reducing the ICER following a negative NICE HTA review (Figure 6.3)

6.5.2 Some Countries Are More Resistant Than Others

- Denmark, France, and Spain are less appreciative of MAAs.

- Denmark
  - Resistance to the administrative burden of access schemes
- France
  - Reliance on payback agreements

- Resistance to the concept of risk sharing
- Little interest in outcome-based agreements
- Lack of transparency
- CED schemes are increasingly implemented
- Spain
  - Uses price/volume agreements, discounts, and rebates
  - Risk-sharing agreements at the regional level have been piloted in some hospitals; details are kept confidential
  - Regional authority of Catalonia initiated the first official risk-sharing agreement in oncology

6.5.3 Regional MAAs Growth in Europe

- MAA can be implemented in different ways, in different regions within the same country, with different levels of autonomy.
- The increasing budget pressure and the lack of regional market access culture in the industry hinder the regional market access for pharmaceuticals.
- To date, the industry struggles to grasp the nuances of market access at a regional level.
- Although at a national level the focus is put on the scientific data, regional payers are more often concerned with the budget impact.

The types of MAAs and their implementation vary across countries but also across regions within the same country and between the central NHS and local payers in the same country.

The increasing budget pressure at the regional level has contributed to the creation of a new major hurdle for pharmaceuticals' market access.

The lack of market access culture in the industry's field teams and the large number of regions has led to a poor management of regional market access. There is a trend to structure the regional management of MA from

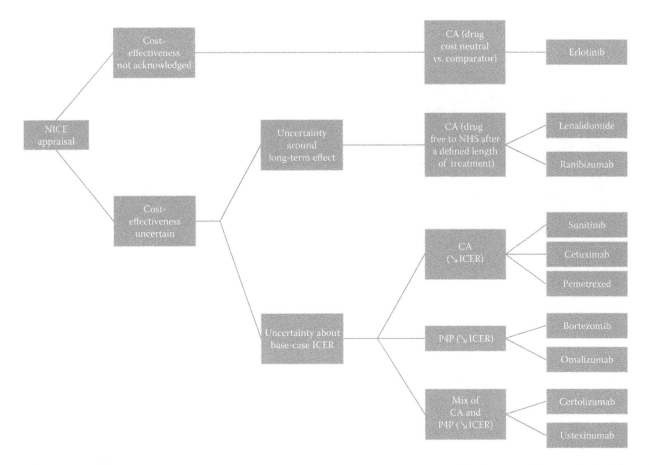

FIGURE 6.3 Classification of PASs according to the UK NICE's consideration of the cost-effectiveness evidence and the design of the PAS. ICER, incremental cost-effectiveness ratio. (From S. Jarosławski and M. Toumi, Design of patient access schemes in the UK: Influence of health technology assessment by the National Institute for Health and Clinical Excellence, *Appl Health Econ Health Policy*, 9(4), 209-215, 2011.)

pragmatic prioritization and segmentation, like in Italy, to a very sophisticated segmentation process, like in the United Kingdom.

In practice, the balance of power and the level of autonomy are very different from one country to another. In general, the federal/central government sets a vision, whereas the local/regional authorities will decide on implementation.

Payers at a national level decide on the following:
- Price and reimbursement
- National MAA
- Recommendation

Payers at a regional level manage the following:
- Entry date
- Local MAA
- Prescription restrictions
- Local recommendation

Until now, the industry has not yet succeeded in addressing regional hurdles in a rational and systematic way as done for national bodies. This is due to numerous reasons:

- Complex and very atomized market
- Need for tailor-made solutions

- Rules are not transparent
- Divergent requirements within the same country

At the national level, decision makers will focus on the scientific evaluation, whereas at the regional level, they are mainly concerned with the budgetary impact. This explains why CAs are more frequent in the regionalized markets of Italy and Spain.

6.5.3.1 Formularies in Italy Are Subject to Regional Influence

MAAs for oncology drugs are negotiated by the AIFA nationally, and the concerned companies are managed, through the "Onco registry," by hospitals (clinicians and hospital pharmacists). Even though regions are not allowed to negotiate MAAs, most regions govern MA through regional formularies, tenders, and actions on prescribing behavior. Some regions are simply putting pressure on hospitals to apply MAAs and ask for reimbursement to the relevant company for nonresponders.

In Italy, regions are responsible for any deficit they incur on health-care expenditure (Figure 6.4). They decide the overall structure of their (regional) health-care system (quasi-market mechanisms, accreditation, cost containment actions, including pharmaceuticals), provided that they give the "essential level of health care" determined by the central government.[9] Thus, the

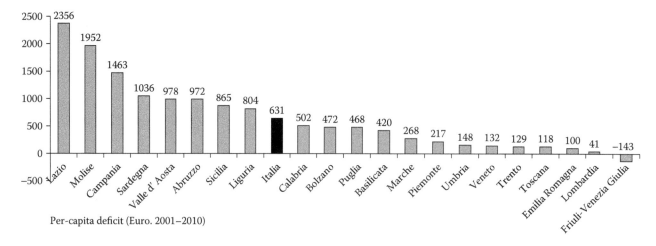

FIGURE 6.4 Italian regions and deficits in health-care budget.

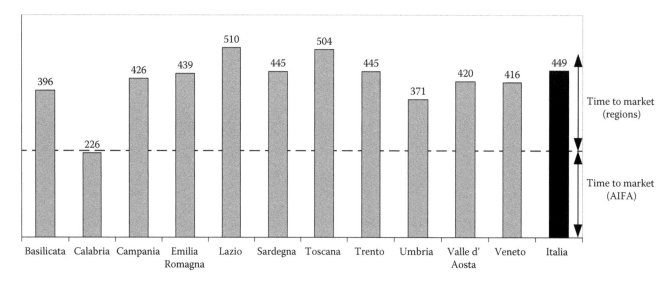

FIGURE 6.5 Regional differences in time to market for new drugs in Italy. (From C. Jommi, *Central and Regional Policies Affecting Drugs Market Access in Italy*, Bocconi University, Milan, Italy, 2010.)

regions are more sensitive to the impact of drugs on the health-care system as they are responsible for budget on drugs used in hospital setting (2.4% of health budget) (Figure 6.5).

In practice, short-term perspective and comparative analysis, based on the cost per daily defined dose (DDD)/therapeutic cycle, prevail.

A study in Italy in 2010[10] highlighted the variability of patients' access to oncology products among Italian regions (IRs).

- The percentage of patient access by IRs, among the 14 oncology products, ranged from 50% to 85.7%, and the mean delay from patient access was 5.3 months.
- Not all oncology products authorized by the AIFA have subsequently been released in every IR.
- "The statistically significant relationship between economic predictors and time to regional access is consistent with the problem of sustainability of pharmaceutical expenditure, especially relevant in the case of cancer therapies, where costs of care have generated some concerns."[11]

- Regional formularies have gained a more dominant role compared to national formularies. This represents a further barrier to patients' access to drugs in Italy.
- Results from this study show that a much faster patient access for oncology drugs was achieved with products receiving a market authorization combined with a risk-sharing agreement.
- Italy is one of the first countries where "risk-sharing agreements" for oncology products have been introduced (i.e., from 2006), and the results of monitoring registry are publicly available (Table 6.6).[12]

6.5.3.2 Regional MAA in Sweden

In Sweden, funding of health care is local (county council) and MAAs are negotiated locally.

- Less visible internationally
- Less transparent
- An increase in MAAs implies more requirements to negotiate and administer

TABLE 6.6
Percentage of Patient Access by IR 10

Italian Regions	Number of Available Products	Percentage of Patient Access[a]	Percentage of Stable Patient Access[b]
Campania	12	85.7	85.7
Friuli-Venezia Giulia	12	85.7	85.7
Lombardia	12	85.7	78.6
Marche	12	85.7	78.6
Piemonte	12	85.7	85.7
Toscana	12	85.7	71.4
Veneto	12	85.7	71.4
Abruzzo	11	78.6	71.4
Lazio	11	78.6	78.6
Liguria	11	78.6	71.4
Puglia	11	78.6	78.6
Calabria	10	71.4	57.1
Emilia Romagna	10	71.4	71.4
Sicilia	10	71.4	71.4
Umbria	10	71.4	64.3
Basilicata	9	64.3	57.1
Provincia autonoma di Bolzano	9	64.3	42.9
Provincia autonoma di Trento	8	57.1	42.9
Sardegna	8	57.1	50.0
Molise	7	50.0	35.7
Valle d'Aosta	7	50.0	21.4

Source: C. Jommi, *Central and Regional Policies Affecting Drugs Market Access in Italy*, Bocconi University, Milan, Italy, 2010.

[a] Computed with respect to the 14 oncology products with MA in Italy.

[b] The purchase date has been defined according to the handling month in which the volume corresponded to at least 20th percentile of the overall volume of the given product.

For example, in case of Roche's bevacizumab:

- Stockholm County Council agreed in April 2008 that if patients with advanced cancer exceeded an accumulated dose of 10,000 mg of bevacizumab, the additional costs will be covered by the company.
- The scheme has been extended, and the other regions in Sweden have been offered similar schemes by the firm.

Local MAAs are growing in other countries as well:

- Spanish regions
- UK General Practitioner practice units after Primary Care Trust reforms

6.6 BEST PRACTICE OF MAAs

- Establishing a general implementation guide for MAA requires identifying the rationale behind the MAA typology, the implementation process, and the evaluation of these agreements.

All MAAs do not follow the same process, and they are implemented differently according to various parameters (underlying motivations for the manufacturer and the payer, the type of drug, the country, etc.).

Even though each agreement is unique, it is possible to define a general implementation guide for MAA that consists in identifying the rationale behind the MAA structure, the implementation process, and the evaluation of these agreements.

6.6.1 THE RATIONALE BEHIND MAAs

6.6.1.1 When Should MAAs Be Considered?

- MAAs are relevant within the context of an MA strategy for a new drug.

Table 6.7 resumes the factors which indicate that an MAA (P4P or CED) should be considered within an MA strategy for a drug. The right-hand column contains the payer's point of view and therefore defines what evidence should be provided. CAs are typically simple rebates or discounts with low-administrative burden and are not included in the table.

TABLE 6.7
Rationale behind MAAs: Factors to Consider

Factors Favoring Risk Sharing	Points to Consider
1. Disease with high unmet needs	• Outcomes, side-effects, administration
2. Disease is severe or life threatening	• Mortality rates, burden of disease
3. Uncertainty related to patient population	• Identification of patients, similar indications, and degree of off-label use in therapy area
4. An objective and measurable outcome, ideally with biomarkers	• Primary versus proxy outcome
5. An innovative treatment with uncertainty over expected risks and benefits	• New drug class, new treatment approach, new endpoint
6. High cost per treatment	• Total potential population (on/off-label)
7. Large budget impact	• High-cost therapy
8. Strong political support and/or patient demand for drug access	• Political priority
	• Well-organized patient groups

6.6.2 THE IMPLEMENTATION PROCESS

6.6.2.1 Requirements for Implementing MAAs

- Plausible outcomes, acceptable costs, realistic time horizon, clear funding arrangements, and allocation of responsibility for data collection and analysis as well as potential discounts are the prerequisites for an operable MAA.
- The MAA scheme will depend on the nature and level of expected outcome.
- Administrative arrangements, future adjustments, and exit plans must also be taken into consideration.

In order to conclude MAA agreements, a number of elements must be in place. The task force for performance-based risk-sharing arrangements (PBRSA) identified the requirements for their implementation:[13]

- Measuring appropriate outcomes that should be clinically robust, plausible, appropriate, and possible to monitor.
- *Acceptable costs*: The costs of the schemes to the health-care system should be proportionate to the potential gains.
- *Realistic time horizon*: It should have clear target date. According to Hutton et al., an MAA longer than 3 years will be irrelevant in the face of changing clinical practice and technological advancement.[14]
  - It is important that all relevant new data can be accomplished within a realistic period.
- Funding arrangements must be clear whether it is the manufacturer's or payer's responsibility.
- Define how the responsibility for data collection and analysis is allocated.
- Define the process of analyzing and reviewing the evidence in order to make a revised decision on price, revenue, or coverage.

- Define if any rebates or discounts will be paid during the course of the scheme, based on provisional results for instance.

The design of the scheme will depend on the following:

- Nature of the expected outcome
- Expected level of results

In general, schemes must not involve large administrative burden and must take into consideration possible changes that may impact the agreement, such as changes in therapeutic guidelines. Also, exit strategies must be considered, for example, for a case in which the new drug turns out to be less effective, which would lead to funding withdrawal during the lifetime of the scheme.[5]

6.6.2.2 Challenges in MAAs Implementation

- The cost of data collection and the transferability of the data, as well as the payers' analytical skills, are regarded as important challenges to the implementation of MAAs.

Apart from the fact that both parties must agree on the design of MAAs, there are some practical difficulties for MAA implementation:

- There is a question concerning the transferability of certain results of a given agreement, as cost data or quality of life (QoL) data may be more country/region dependent than clinical results.
- Collection of data requires suitable information systems and careful sampling to avoid biases.
- Analytical expertise and good information system need to be available for the payer in order to design the protocol and monitor it over time.

6.6.3 EVALUATION

- MAAs are evaluated in consonance with their own parameters and bearing in mind their cost of implementation and conduction.
- Success of MAAs relies on many factors, such as the measure of the outcomes, reduction of uncertainty, and compliance with the budget.
- Overall, there is little information available in the literature on the evaluation of MAAs. This is expected to change as these agreements become more transparent.

In theory, evaluation of MAAs should include the overall costs involved with implementing and conducting the schemes, as well as the health outcomes.

The Task Force Report[13] on PBRSA indicates that evaluation should rely on process indicators for the scheme's success: "It will be an important part of the design of any PBRSA scheme to define the metrics by which the success of that scheme can be assessed."

They identified the following questions to seize the process indicators of success:

- Were the intended outcome measures collected?
- Was uncertainty in associated parameter estimation reduced for the outcomes that were the focus of the scheme?
- Did the scheme keep within the budget?
- Was the integrity of the design/estimation maintained?
- Did the governance arrangements work?
- Did the process to underpin a decision with further evidence work?

"The appropriate decision-making will require the ability to show that the agreed outcome adjustments were made to guarantee the cost-effectiveness of the intervention."[13]

Until now, little has been published on MAA evaluation of schemes and results:

- A systematic literature review conducted[15] in 2011 found that the studies referring to the UK multiple sclerosis (MS) MAA (40% of all studies) reported a few costs, whereas other studies included qualitative discussions of costs and benefits, without evaluating the overall economic impact of the scheme.
- Other articles[16,17] were published regarding early results in monitoring the study in MS. They reported that "baseline characteristics and a small but statistically significant progression of disease were similar to those reported in previous pivotal studies." However,

a conclusion of both articles is that it is too early to conclude on the cost-effectiveness of the MS disease-modifying drugs.

It is important to note that not all proposed schemes will be accepted by payers, even though the majority is. It will depend on "the payer's needs in terms of clinical effectiveness, cost-effectiveness, and/or budget impact of the needs of the manufacturer in terms of pricing, access and revenues."[6] Among the schemes that had sufficient details available for evaluation (46 out of 53), 72% gained unlimited access to the market and 28% gained limited access to the market (research participation, cost-effective subgroups, and prior authorizations).

Hopefully, more publications on evaluation of MAAs will be found as these agreements become more transparent.

6.7 IMPACT OF MAAs ON PRODUCT UPTAKE

- Patients and pharmaceutical manufacturers have different expectations and draw different benefits from MAAs.

The main rationale for manufacturers to enter into an MAA is to ensure profitability, optimize sales volumes, and market share of a specific drug while avoiding the negative impact of international reference pricing. Patients expect that those agreements will allow them to obtain access to expensive innovative products.

The adoption of such schemes is not without the impact on the market uptake, which may be either positive or negative depending on the clinical or financial outcomes.

The direct and indirect impacts on sales and implications of MAA will be presented in this section based on actual cases.

6.7.1 EXAMPLE OF MS MAA IN THE UNITED KINGDOM: IMPACT ON SALES

- The PAS had a perceptible negative impact on MS patients' access to drugs in the United Kingdom.

- UK sales of MS disease-modifying drugs were the lowest in EU5 (top 5 European Union markets) as they were impacted by the PAS (Section 8.1) (Figure 6.6).
- In 2004, the uptake was very slow; 20% of eligible patients were still waiting to see a specialist doctor. Only 8% of MS patients received the disease-modifying treatment.
- In 2007, there were 11.5% of MS patients receiving treatment in the United Kingdom, 35% in Western Europe, and 50% in the United States.
- The MAAs for MS biologic drugs did not improve British patient's access (Table 6.8).

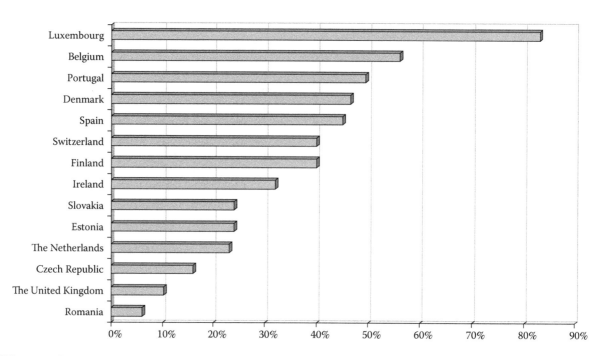

FIGURE 6.6 MS-estimated proportion of patients on treatment in 2008. (From C. Jommi, *Central and Regional Policies Affecting Drugs Market Access in Italy*, Bocconi University, Milan, Italy, 2010.)

TABLE 6.8

Cumulative Sales of Disease-Modifying Drugs for Multiple Sclerosis from Drugs' Launch to 2008

Country	Cumulated Sales ($ M)
Germany	635
France	284
Italy	219
Spain	174.5
UK	76.5

Source: Courtesy of IMS sales.

6.7.2 Example of Bevacizumab's Uptake in Metastatic Colorectal Cancer across EU Countries

- The usage rates for the same medicine vary widely among countries with respect to the funding practice and the value assessment methods.

Utilization rates of bevacizumab in metastatic colorectal cancer (mCRC) varied across the European Union countries, depending on HTA and funding schemes (Figure 6.7):

- Funding on top of the disease-related group (DRG) and lower price in France resulted in higher utilization rate.
- The same type of funding at a higher price in Germany resulted in relatively low usage of bevacizumab.
- MAAs based on payment by performance in Italy allowed high utilization rates and high price (however, recent results of the scheme suggest price revision with potential 30% price discounts).
- The lack of NICE's recommendation in the United Kingdom caused virtually no use.
- Bevacizumab uptake in Italy would have been much slower without a risk-sharing scheme for the drug.
- In France, early uptake (prior to European Medicines Agency [EMA] license) was possible due to temporary authorizations for usage (ATU) (pre-registration usage authorization) and unrestricted funding on top of DRGs. Rather than MAA, early access due to ATU and funding on top of DRG increased the sales of the drug in France.

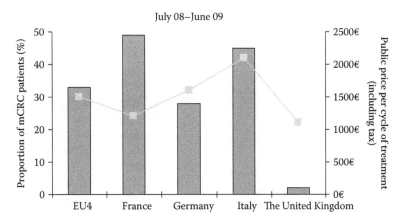

FIGURE 6.7 Utilization rates of bevacizumab in mCRC varied across the European Union, depending on HTA and funding schemes. (From M. Toumi et al., Influence of health technology assessments on utilization of Bevacizumab in Europe, PCN140, *ISPOR 15th Annual International Meeting*, Atlanta, GA, 2010.)

6.7.3 ETANERCEPT MAA IN GERMANY

- As part of a risk-sharing agreement, Wyeth Pharmaceuticals (Dallas, TX) agreed to provide compliance support to patients on etanercept. This turned out to be a successful experience and many such agreements have been implemented since by other companies and payers.

An MAA for rheumatoid arthritis was engaged in Germany for etanercept between Wyeth Pharmaceuticals and the third largest statutory health insurance (Taunus BKK) fund in Germany in 2008.[18] Wyeth would agree to fund and provide compliance support to patients taking etanercept which in turn would significantly improve the treatment's effectiveness. Indeed, being an injectable drug, etanercept effectiveness would highly depend on patients' compliance. In fact, one-third of patients discontinue the treatment within the first 3 months due partly to lack of compliance.

The company offered telephone-line support service, homecare visits by qualified nurses, and the promotion of regular patient communications about treatment, as well as tips on how to self-inject and the importance of maintaining therapy.

This scheme allowed patients to become more compliant to the treatment and to remain on treatment for a longer time, thus experiencing a more effective treatment.

"Since the scheme, etanercept has shown a more positive sales trend than its competitors and Wyeth has expanded the deal to over 100 other German sick funds, with some even eliminating co-payments on the drug to improve uptake."[18]

6.8 SOME SPECIFIC CASE STUDIES

6.8.1 EXAMPLE OF CED IN THE UNITED KINGDOM: USE OF β-INTERFERONS AND GLATIRAMERE FOR THE TREATMENT OF MS IN THE UNITED KINGDOM

- Under external pressure, the NICE reconciled themselves to fund β-interferons for the treatment of MS. To this purpose, they established a 20-year scheme aiming to demonstrate the outcome of the product in terms of cost per quality adjusted life year (QALY).
- This scheme was proven to be inconclusive, and it resulted in restricted patient access as well as an unsatisfactory amount of collected data.
- The scheme was heavily criticized for its "flawed, scant, and speculative model," its length, as well as the insufficient funds allocated to it.

This agreement between the UK national payer and the pharmaceutical industry was an early example of the CED concept based on the demonstration of a drug's effectiveness in real life.[3] The company was asked to demonstrate its claims of the impact of the treatment on QoL and its efficiency, measured by an incremental cost-effectiveness ratio per QALY.

In its initial appraisal, NICE refused to fund β-interferons for the treatment of MS on clinical and cost-effectiveness grounds (the calculated cost/QALY was £42,000–£98,000 over 20 years and would rise to a maximum of £780,000/QALY over 5 years).[3]

Under political and patient associations' pressure, in 2002 the government established a scheme with four MS drug manufacturers, where a cohort of approximately 10,000 patients would be followed for over 10 years with the cost of drugs reduced or refunds given to the payer if the cost/QALY over an envisaged 20-year horizon was over £35,000/QALY, that is, to fund a maximum value of £35,000/QALY or less.[5]

Patients would be followed using the Kurtzke Expanded Disability Status Scale (EDSS), which was the same outcome measure as used in the trials.

The initial assessment, published in 2009, highlighted the important issues concerning the study's methodology, insisting on the need to have longer term follow-ups before securing meaningful results. The experimental CED for MS drugs in the United Kingdom was inconclusive at 7 years from its launch.

In this case, the so-called PAS restricted access and led to much lower usage rates than the drugs would have otherwise achieved with positive recommendation from the NICE. In fact, the United Kingdom ranked 13 out of 14 high-income countries for use of new disease modifying drugs (DMDs) in 2010, despite the PAS being in place since 2002.[7]

6.8.1.1 Performance of MAAs

The β-interferons scheme in the United Kingdom was heavily criticized for a number of reasons, which need to be taken into consideration by payers for future scheme evaluations. These include the following:[5]

- The model
  - Flaws in the actual model (difficulties in fully mapping out the quality of life and the natural history of MS to the trial outcomes, which were based on changes in EDSS scores)
  - Concerns that the model was heavily influenced by assumptions about future discounting and did not account, for example, for the cost of azathioprine
  - Concerns that the model did not appear to fully account for patients discontinuing treatment early because of side effects
- The length of follow-up
  - Concerns that within 10 years the β-interferons and glatiramer acetate may have been replaced by newer drugs questioning the whole rationale behind the scheme
- Funding and administration support
  - Primary care trusts generally did not receive any additional funding to cover the cost of these drugs
  - Hospitals also did not receive additional funding for more extensive follow-up consultations and for completing the necessary forms reducing their involvement in practice
  - Concerns with the necessary infrastructure required including specialist nurses, as well as concerns where the costs of the additional administrative burden would come from

6.8.2 EXAMPLES OF CEDs IN FRANCE

- The French HTA agency (*Comité Economique des Produits de Santé* [CEPS]) allowed an escrow MAA for long-acting injectable risperidone, stating that the drug would be granted a premium list price, but the company would be paid based on the price of the cheapest comparator until the outcome is proven. In conclusion of the scheme, the premium list price was upheld.
- The CEPS and the Transparency Committee conducted a 2-year observational study to investigate the glitazones manufacturer's claim of a superior real-life efficacy. The study was conclusive and the benefit of the drug was established.
- Omalizumab's launch in France was conditional upon conducting a 2-year study that demonstrates a positive impact in real life. The drug failed to meet the authorities' expectations.

We describe three requests of the *French Haute Autorité de Santé* (HAS) of real-life comparative studies. These were expected to reduce uncertainty about drugs' real-life performance and enable final pricing.

- In 2005, the French pricing committee CEPS asked the manufacturer of the long-acting injectable risperidone to perform a 1-year study that should evidence reduction in the rate of hospitalizations for patients treated with this drug compared to other antipsychotics (to be designed under supervision of the Ministry of Health).[19,20] This escrow MAA assumed that although the drug would be granted a premium list price (approximately 15-fold premium vs. generic injectable long acting antipsychotics, and almost 60% vs. oral risperidone from Janssen–Cilag), the company would be paid based on the price of the cheapest comparator. Then, the difference would be deposited as public funds in Caisse des Dépôts et Consignations until results from the study were available. Should the results evidence reduction in hospitalization rate, money would be transferred to the company. Otherwise, social security services would receive the funds.
- Another MAA requested by CEPS and the Transparency Commission in 2004 was a real-life use study for glitazones (pioglitazone and rosiglitazone) in type 2 diabetes. This 2-year observational study intended to develop evidence that would support or invalidate the manufacturer's claim of a superior real-life efficacy (time to introduction of an add-on therapy) compared to what was previously observed in clinical trials.

- In 2006, omalizumab's launch in France was conditional upon conducting a study that demonstrates a positive impact in real life. The prospective study was conducted on a cohort of 1,000 patients that were followed for 2 years. The main criterion was the onset of a severe asthma exacerbation defined by at least one of the following events: hospitalization, emergency room (ER) visit, oral corticoid prescription, and increase in oral corticoid dose by at least 20 mg equivalent prednisone.

6.8.2.1 Performance of MAAs

- Five years after the initial HAS ruling of minor improvement of clinical benefit (*Amélioration du Service Médical Rendu* [ASMR] IV) for long-acting injectable risperidone, the requested study provided evidence that in a cohort of over 1,600 patients followed for 1 year, patients treated with the drug had a lower relative risk of hospitalization compared to other antipsychotics.[19] Thus, the premium list price was maintained.
- Similarly, results of omalizumab's study showed real-life benefit with a better control of persistent severe asthma in terms of prescriptions and hospitalizations and/or ER visits (44% decrease).
- On the contrary, the observational study for rosiglitazone (AVANCE) did not support the manufacturers' claims of a superior real-life efficacy, as the study largely repeated the efficacy data, which had been shown in clinical trials.[20,21] Consequently, the pricing committee CEPS cut the drug's price by 30% and requested rebates for the drugs that were already purchased. The reimbursement level also dropped from 65% to 35%. The actual amounts of the rebates were not published by HAS, however.

6.8.3 Examples of CEDs in Sweden

The Swedish TLV uses cost-effectiveness analysis to inform its decision making; thus, it will not recommend the reimbursement of drugs that show uncertain or high ICER value (depending on a specific disease threshold).

These two agreements provided the necessary data to assign a definite reimbursement status for the drugs:

- Levodopa/carbidopa intestinal gel went through a temporary (5-year-long) MAA which aimed at generating real-life evidence that would allow a reduction of the value of ICER and uncertainty

around it and help achieve the final pricing path and reimbursement in Sweden.[22,23] Briefly, at the time of initial manufacturer submission, although the product was granted provisory reimbursement at a premium price, the follow-up studies and additional cost-effectiveness analyses allowed the TLV to give a positive final reimbursement decision premium price.

- Rimonabant, a drug in the treatment of obesity, was granted provisional reimbursement status for 2 years. At the end, TLV's final decision was conditional to provide additional data that would evidence cost-effectiveness of the drug as well as long-term effects in real practice.[23–25]

6.8.3.1 Performance of MAAs

These two CEDs enabled the payer to collect evidence that led to reevaluation of the real-life cost-effectiveness of concerned drugs, and final reimbursement and pricing decisions were delivered. Following these decisions, the drugs were financed without employing any further MAA.

6.8.4 Example of MAAs in Italy

- Italy uses a wide variety of MAAs, most of which are disguised price cuts with haphazard data collection processes without precision about the desired/assessed health outcome.
- The CRONOS project monitored real-life effects of drugs in Alzheimer's disease (AD) patients and gave rise to individual reimbursement decisions.
- The AIFA resolved to inflict price cuts upon expensive cancer drugs, breaching agreed oncology MAAs.

Nineteen various MAAs were identified in Italy until 2011, but there were no detailed technology appraisals available on the AIFA's website.

Overall, there were 12 MAAs for oncology drugs: erlotinib (2006), sunitinib (2006), sorafenib (in advanced renal cell carcinoma in 2006 and liver cancer in 2008), dasatinib (2007), bevacizumab (2008), lenalidomide (2008), temsirolimus (2008), bortezomib (2009), cetuximab (2009), lapatinib (2009), panitumumab (2009), and trabectedin (2009).[26]

- The MAA involved fixed discounts (from the list price) and/or paybacks for nonresponding patients (100% or 50% of the drug's cost, all on a per-patient basis).
- Although safety and efficacy of the drugs were monitored in patient registries, it is noteworthy that those MAAs did not seek to answer uncertainty about a clearly specified health outcome, and data collection

in the registries was not systematic with a high potential for various bias.[27]

- For example, it was estimated that in some regions of the country, only 50% of patients were covered by the registries and no process had been put in place to ensure an unbiased selection of patients, which might have led to doctors feeding records only for patients for which less administrative burden was expected.[28]

In contrast, the CRONOS project launched by the AIFA to evaluate the real-life effectiveness of AD drugs (donepezil, rivastigmine, and galantamine) assumed collection and analysis of well-defined health outcomes from a cohort of patients. It was carried out in a nationally representative sample of patients with AD over a period of 2 years, and the public insurer reimbursed medicines only in patients who responded at 4 months of treatment (while the cost for nonresponders was covered by manufacturers).[5,29]

6.8.4.1 Performance of MAAs

- The Italian AIFA announced that following an analysis of patients' registries, which were a part of P4P and CA agreements for expensive cancer drugs, it would reduce their list price by 30%–40% in 2011.[30] Although scheduled price revision was assumed at the launch of these P4P schemes, they had not been designed to answer uncertainty about the (cost-) effectiveness of the drugs, and it is unlikely that the registry data had the sufficient quality to provide more robust estimates than those available at drugs' launch.
- However, the Italian CED for AD drugs (CRONOS), provided new real-life effectiveness data and allowed AIFA to reimburse these medicines, with some restrictions with respect to diagnosis and continuation of treatment and prescription limited to specialist physicians.[31]

6.8.5 Financing of MAA Drugs

Noteworthy, the payers have a diversity of approaches toward how to sponsor the provision of the drug for patients during its testing CED period.[3] In France, drugs in one CED were financed by the payer at a premium price, but in case of unfavorable results, the difference between the cost of a cheaper comparator and the premium price was returned to the payer. In Sweden, the drugs were financed at a premium price and in Italy at a premium price with payment limited to patients who responded to treatment (for the CRONOS MAA). In the United Kingdom, the MS disease-modifying drugs were financed at a premium price, but the scheme assumed sliding reduction of the price as soon as unfavorable interim evidence from the CED became available.

6.9 OVERVIEW OF MAA TRENDS IN OTHER COUNTRIES

- MAAs are often confidential and only little information is published.
- They hold a certain amount of confidentiality, and it is difficult to determine the exact number and details of the agreements in place in various countries.

Most of the information presented in Section 6.9.1 was extracted from the Jaime Espin et al. report[32] prepared for European countries and based on results from a survey, literature review, conferences, and other gray literature for the European commission. In other countries such as Latin American countries and Australia, information was extracted from other published literature.

6.9.1 MAA Is a Growing Phenomenon in Various Countries

- In the recent past, the practice of MAAs has been increasing and expanding to new countries.

MAAs have been increasing in the recent years. The authors found that between January 2010 and June 2011, there were 45 new agreements, nearly double the total of MAAs for 2009. The countries where new agreements were mostly found were the United Kingdom (40%). Italy and Australia were other important markets, and a significant number of new countries are beginning to see MAAs such as New Zealand, Belgium, Poland, and Hungary.[1]

6.9.2 MAAs in Australia

- The use of MAAs is rapidly expanding in Australia and has covered most of the recently launched drugs.
- No information has been made public on the type of MAAs used except for sapropterin hydrochloride.

Australia is one of the countries in which MAAs have been rapidly expanding recently.[33]

Indeed, up to June 30, 2012, there were 76 deeds of agreement in place or in development.[33,34]

In 2012, 15 new drugs were listed, of which 8 received a recommendation for an MAA by the Pharmaceutical Benefits

Advisory Committee (PBAC). It was noted that seven of the eight were recommended fully or in part on an incremental cost-effectiveness analysis (CEA) basis, suggesting that "it is highly likely that a new medicine recommended by the PBAC on an incremental CEA basis will be associated with a RSA."

Among the eight drugs, three of them were indicated in oncology, two were destined for hepatitis C, and one was recommended for an orphan indication.[33]

No detailed information was given on the type of MAA recommended, except for sapropterin hydrochloride (orphan drug in phenylketonuria), where it was specified that the PBAC recommended an agreement with a 100% rebate in case the number of patients with tetrahydrobiopterin deficiency should exceed 20. There are only 12 patients treated with sapropterin in Australia.[33]

6.9.3 MAAs in Latin America

- An HTA network was launched in Latin America to remedy the fragmentation of the health-care systems and the lack of resources and expertise.
- Cuba, Brazil, and Mexico have published HTA guidelines; however, they remain focused on the efficacy and safety more than the cost.
- Latin American countries have no experience with P4P agreements. Only Brazil has shown a recent, but still unconfirmed, interest in risk-sharing agreements.

HTA has developed recently in Latin American countries when the Organizacion Panamericana del Salud (Pan-American Health Organization) created a new HTA network in the Americas (*Red de Evaluación de Tecnologías Sanitarias*). This was done to address the fragmentation of the health-care systems in the region and the lack of resources and expertise for some countries.

Cuba, Brazil, and Mexico have recently developed and published methodological guidelines in economic evaluation, indicating that there is a growing interest in evaluating health-related products, drugs, and technologies used by the population.[35,36] However, until now in these countries, decisions regarding drug approval and funding are made on the basis of efficacy and safety and not cost.

So far, these countries have no experience in P4P contracts or conditional reimbursement. Only in Brazil have discussions regarding these programs started. However, according to the current Brazilian regulation, risk-sharing negotiations are not allowed. This might potentially change in the next few years.

6.9.4 India: A Different MA Pathway

- MAAs are difficult to implement in India, although they have been used increasingly in other Asian countries. However, most of them are regarded as marketing schemes rather than genuine MAAs.
- The hurdles to establishing MAAs in India are essentially the bureaucracy and the informal economy and the sometimes unethical medical practice.

A recent article[37] has explored the opportunity to implement MAAs in India through desk and primary research interviewing key stakeholders. The author concluded that it is difficult to implement such agreements in India.

Results from the desk research and the primary research showed that

- Recent developments in India, including the introduction of compulsory licensing for sorafenib may generate greater pressure on the pharmaceutical industry to reduce price.
- Physicians consider most of the current schemes, with a few exceptions, as marketing schemes rather than genuine patient assistance programs. Only those schemes implemented closely with locally based foundations/cancer institutes are considered by physicians to be true assistance programs.
- Existing hurdles for MAAs that were identified are as follows:
  - Bureaucracy and the informal economy (which creates difficulties with means testing) are the two most important reasons for the schemes not "taking off" (quoted by the majority of respondents).
  - General ignorance about the potential of such schemes and cynicism surrounding them are also a deterrent to scheme success.
  - Unethical medical practice, knowledge gap, and a lack of streamlined infrastructure for scheme delivery are also thought to be important factors.

Overall, the authors concluded that "Establishing cost-sharing schemes in India remains a challenge because of the fractured infrastructure, porous supply-chain and unethical medical practice."[37]

6.9.5 MA in South Korea

- South Korea has recently implemented its HTA process. Its approach appears to be driven essentially by price discounts and price volume agreements.

MA for oncological drugs in South Korea has been recently compared to that of Australia and the United Kingdom through the review of HTAs.[38] These showed that the South Korean HTA process is new, not well defined, and that the pharmacoeconomic assessment is still evolving.

Although cost-effectiveness is the main driver for a positive recommendation, South Korea's approach appears to be less rigorous and mainly aimed at financial agreement such as price discounts and price/volume agreements, which play a major role in determining reimbursement.[38]

6.10 PERSPECTIVES

6.10.1 MAA Is a Growing Trend and Is Shifting toward Conditional Access

- There are at least 10 new MAAs implemented every year. Very little information about these agreements is ever disclosed.
- Although CA and P4P provide for savings on the individual payer's level, their impact on external reference pricing is expected to be detrimental.
- P4P agreements are difficult to implement and will probably decline over time, whereas CEDs are likely to develop as a reference tool for managing the payer's uncertainty.
- In the future, MA is expected to be achieved through conditional approval and conditional access.

Since 2003, there has been a fast growth of MAA, with approximately 10 new MAAs established every year (Sullivan 2009). There are probably more MAAs put in place every year, as most are and not publicly disclosed.

However, it remains unknown how many new molecules are financed through MAA every year, as very little information is made publicly available.

Whereas CA and P4P have the potential to incur savings to individual national payers, care should be taken that the real cost of purchasing drugs in MAAs, rather than their list price, is used for international reference pricing in Europe.[3]

P4P schemes based on outcomes in individual patients have proven difficult to implement and manage because of their administrative burden on hospitals, which already suffer from lack of resources.

On the contrary, CED concept seems to grow in Nordic countries, the Netherlands, France, and others. It represents a powerful tool, which enables the reduction of uncertainty about a drug's real-life performance. However, every question or issue raised could lead to a specific study that requires expertise and a mutual agreement from the payer and manufacture on the study conditions. This procedure allows a rapid MA for new molecules or interventions and gives patients a chance to access innovative treatments earlier. Moreover, it protects the payer's budget, thanks to the escrow facility.

CED is bound to expand in countries with developed HTA competencies as a reference tool for managing the payer's uncertainty. However, P4P seems to be gradually reaching its limits. Finally, CAs are becoming adjustments to list prices which allow manufacturers to avoid price reduction that could affect international reference pricing.

A shift is foreseen from the regulatory approval and MA being achieved through a single decision point toward a decision window through conditional approval and conditional access.

6.10.2 Challenges with MAAs

- In the future, MAAs will be faced with the challenge of defeating the high transaction costs, the administrative burden, and the evaluation flaws, as well as achieving more transparency.

There are a number of challenges that arise when considering health outcomes-based schemes, which limit their long-term impact and viability:

- High transaction costs
- Administrative burden
- Insufficient information systems
- The benefits of MAAs are market dependent, and the transferability from one country to another may be difficult[2]

The future challenges to overcome include the complexity of the schemes, the administrative burden, and the difficulty in evaluating MAAs.

There will be a need for more transparency regarding these schemes in order to understand the attitudes and perceptions of various stakeholders, as well as to evaluate the results and experience with the schemes implemented so far.[6]

6.11 CONCLUSION: MAA, A TEMPORARY SOLUTION?

- In order to adjust to a fluctuating environment, the pharmaceutical industry is shifting toward an insurance business model.
- The approval of new drugs in a risk adverse environment has shifted from a one-off procedure to a system with a decision window.

There are numerous challenges in the current health-care environment: the economic crisis, HTA processes becoming more complex, the criteria for benefit recognition are changing, and the regional hurdles are increasing where budget holders are held responsible for the deficit. All these issues are leading to a number of changes.

6.11.1 A PARADIGM SHIFT IN THE PHARMACEUTICAL INDUSTRY

Originally, "risk-sharing agreements" were nonwritten: in the premarketing phase of a new drug, the company would endorse the development risk, and in the postmarketing phase, it was the payer who would endorse the real-life effectiveness risk.

In the new risk-sharing agreements, it is the company that endorses both pre- and postmarketing risks through P4P or CED.

6.11.2 FROM A DECISION POINT TO A DECISION WINDOW

Increasingly, pricing and reimbursement decisions are no longer made at a fixed time point, but over a period of time during which pharmaceutical companies are expected to provide additional information to reduce any uncertainty around the value of a new product.

It is a general fact that risk adverse attitude has increased in the majority of countries where reimbursement approval has become conditional. Recently, regulators and payers have been dealing with three main facts:

- Conditional approvals
- The lack of efficacy has become considered as an "adverse event," and as such, needs to be reported to payers by prescribers
- Risk management plans that have been instituted to track the lack of efficacy as one of the potential "adverse events"

Therefore, there has been a move from a decision point for regulatory approval, pricing, and MA decisions toward a decision window, which is well illustrated by CED schemes. This allows the regulators and payers to minimize the risk associated with the introduction of a new product. This also contributes to an increased dialog between the regulators and the payers.

6.11.3 WHICH MAAs IN THE FUTURE?

- A viable future for MAAs is subject to particular conditions such as specific targeting and transparency.

Many authors foresee the use of MAAs only in specific situations:

- De Pouvourville[2] considers that MAAs will be viable in a defined setting: for innovative drugs in indications where there is little competition, which target very specific populations to whom it is more feasible to demonstrate a benefit claim
- Likewise, Adamski et al.[5] believe there are only a limited number of situations where risk-sharing schemes should be considered in the future, describing also the key issues that payers need to take into consideration

Situations where MAA could be considered are as follows:

- Explicit and transparent objectives and scope of the scheme
- A novel treatment in a high-priority disease area with an expected net health gain
- New drugs that are effective in priority disease; however, potential concerns with long-term safety remain
- New drugs that could have a substantial beneficial impact on service delivery and patient safety; however, this has been difficult to prove in phase III trials
- Likely health gain that can be determined within a limited time frame
- MAA in a priority disease area "substantially lowers health service costs to enhance reimbursement having factored in all administrative costs"[5]

Among the three main types of MAAs described above:

- CED will probably continue to thrive. It is the most promising type of MAA as it diminishes uncertainty, while moving toward a value-based pricing. Furthermore, it constitutes an incentive for the pharmaceutical companies for research and development.
- CA such as confidential discounts will remain attractive as long as there will be international reference pricing, in order to maintain a premium list price.
- P4P concept may gradually become less popular because it is not supported by a robust methodology that brings an added value (e.g., information on effectiveness). It is rather a form of a disguised discount. It is also difficult to monitor, and it incurs high administrative burden.

The potential impact of MAAs is that they will allow to manage some uncertainties regarding real-life effectiveness and use and fine-tune prices. However, MAAs do not address the issue of affordability and public health priorities. Moreover, MAAs that are difficult to implement may increase burden and costs.

Hence, it is not excluded that MAAs might experience further evolution in the coming years, with the increasing pressure from payers for better health outcomes at fair price and the growing dynamics of interrelated drug markets (e.g., international reference pricing).

REFERENCES

1. Ando G, Reinaud F, Bharath A. Global pharmaceutical risk-sharing agreement trends in 2010 and 2011. September/October 2011 issue (Volume 17, No.6) of ISPOR connections.
2. de Pouvourville G. Risk-sharing agreements for innovative drugs: A new solution to old problems? *Eur J Health Econ* 7(3):155–157, 2006.
3. Jaroslawski S, Toumi M. Market access agreements for pharmaceuticals in Europe: Diversity of approaches and underlying concepts. *BMC Health Serv Res* 11:259, 2011.

4. Towse A, Garrison Jr. LP. Can't get no satisfaction? Will pay for performance help?: Toward an economic framework for understanding performance-based risk-sharing agreements for innovative medical products. *Pharmacoeconomics* 28(2):93–102, 2010.

5. Adamski J et al. Risk sharing arrangements for pharmaceuticals: Potential considerations and recommendations for European payers. *BMC Health Serv Res* 10:153, 2010.

6. Carlson JJ et al. Linking payment to health outcomes: A taxonomy and examination of performance-based reimbursement schemes between healthcare payers and manufacturers. *Health Policy* 96(3):179–190, 2010.

7. Toumi M, Michel M. Define access agreements. Published on Pharmaceutical Market Europe www.pmlive.com/Europe; June 2011.

8. NHS Quality Improvement Scotland. Coverage with evidence development in NHSScotland: discussion paper. Coverage with Evidence Development Workshop, September 19, 2008; Glasgow, Scotland.

9. Otto M. Overcoming the complexity of an increasingly fragmented regional and local market access requirement. Presentation at *ISPOR 14th Annual European Congress*, November 2011.

10. Russo P, Mennini FS, Siviero PD, Rasi G. Time to market and patient access to new oncology products in Italy: A multistep pathway from European context to regional healthcare providers. *Ann Oncol.* doi:10.1093/annonc/mdq097.

11. Sinha G. Expensive cancer drugs with modest benefit ignite debate over solutions. *J Natl Cancer Inst* 100(19):1347–1349, 2008.

12. AIFA. Registro Farmaci Oncologici sottoposti a Monitoraggio-Rapporto Nazionale 2007. Rome: AIFA, 2008; http://antineoplastici. agenziafarmaco.it/Registro_farmaci.pdf.

13. Garrison L, Towse A. Performance-based risk-sharing arrangements—Good practices for design, implementation and evaluation: An ISPOR Task Force Report.

14. Hutton J, Trueman P, Henshall C. Coverage with evidence development: An examination of conceptual and policy issues. *Int J Technol Assess Healthcare* Fall;23(4):425–32, 2007.

15. Barros PP. The simple economics of risk-sharing agreements between the NHS and the pharmaceutical industry. *Health Econ* Apr;20(4):461–470, 2011. doi:10.1002/hec.1603.

16. Boggild M, Palace J, Barton P, Ben-Shlomo Y, Bregenzer T, et al. Multiple sclerosis risk sharing scheme: Two year results of clinical cohort study with historical comparator. *BMJ (Clinical Research Edition)* 339:b4677, 2009.

17. Pickin M, Cooper CL, Chater T, O'Hagan A, Abrams KR, et al. The multiple sclerosis risk sharing scheme monitoring study-early results and lessons for the future. *BMC Neurol* 9:1, 2009.

18. Pugatch M, Healy P, Chu R. Sharing the burden: Could risk-sharing change the way we pay for healthcare? *Great Britain*, October 2010.

19. HAS. Avis de la Commission de la transparence. RISPERDAL CONSTA L.P. Paris; 2010.

20. Renaudin MN. Risk Sharing for Reimbursement and Pricing of Drugs. ISPOR Connections 2010.

21. HAS. Avis de la Commission de la transparence. AVANDIA. 2010.

22. Persson U. European Market Access Environment. The Swedish Experience. In European Market Access University Diploma. Paris; 2010.

23. TLV. Lakemedelsformansnamnden (LFN) Beslut (Decision) 0625/2006. Stockholm; 2008.

24. Persson U, Willis M, Odegaard K. A case study of ex ante, value-based price and reimbursement decision-making: TLV and rimonabant in Sweden. *Eur J Health Econ* 11:195–203, 2010.

25. TLV. Lakemedelsformansnamnden (LFN) Beslut (Decision) 1023/2006. Stockholm; 2006.

26. AIFA. Oncology registries.2010.

27. Gallo PF, Deambrosis P. Pharmaceutical risk-sharing and conditional reimbursement in Italy. In *Central and Eastern European Society of Technology Assessment in Healthcare (CEESTAHC)*; 2008.

28. Jommi C. *Central and Regional Policies Affecting Drugs Market Access in Italy.* Bocconi University, Milan, Italy; 2010.

29. AIFA. Protocollo di monitoraggio dei piani di trattamento farmacologico per la malattia di Alzheimer. Rome; 2000.

30. Jack A. Italy to cut cost of cancer drugs. In Financial Times. London; 2010.

31. AIFA. Progetto Cronos: i risultati dello studio osservazionale. Rome; 2004.

32. Jaime Espín JRaLG. Experiences and Impact of European Risk-Sharing Schemes Focusing on Oncology Medicines. http://www.emi-net.eu.

33. Wonder M. PBAC recommendations and risk sharing arrangements—When does an optional extra become a standard accessory? Blog, Wonderdrug, January 2013.

34. Department of Health and Ageing. Annual Report 2011–2012. Australian government; 2012. Available from: http://www. health.gov.au/internet/main/publishing.nsf/Content/annual-report2011-12. Accessed March 15, 2013.

35. Banda D. Health technology assessment in Latin America and the Caribbean. *Int J Technol Assess Healthcare* 25(Suppl.1):253–254, 2009.

36. ISPOR. Pharmacoeconomic guidelines around the world. Available from: http://www.ispor.org/peguidelines/index.asp. Accessed February 6, 2012.

37. Kirpekar S. Failure for cost-sharing schemes to take off in India: What can be the access solution? *Value Health* 15(7):A628, 2012.

38. Saraf S, Akpinar P. Similar HTA, Different Access Outcome? Comparison of Orphan Oncology Drug Assessment in S.Korea, Australia and the UK. *ISPOR 5th Asia-Pacific Conference*, September 2–4, 2012, Taipei, Taiwan. PHP78, Poster Session I.

39. Toumi M. et al. Influence of health technology assessments on utilization of Bevacizumab in Europe, PCN140, *ISPOR 15th Annual International Meeting*, Atlanta, GA, 2010.

40. Sullivan S. D. Pharmaceutical Outcomes Research and Policy Program, University of Washington, Seattle, WA, 2009.

7 External Reference Pricing

7.1 DEFINITION OF EXTERNAL REFERENCE PRICING

Among the member states (MS) of the European Union (EU), each country is free to implement its own national pricing and reimbursement policies as well as to endorse regional agreements, provided that these abide by the Transparency Directive.[1] Therefore, there is a wide variety of pharmaceutical pricing regulations, especially given the historic, political, legal, and economic diversity across countries combined with the variety of health-care systems in terms of funding and organization. However, these regulations are not fully responsible for price differentials between the EU MS on both patent and off-patent markets.[2–4] Despite the large number of cost containment measures ratified by the MS of the EU throughout the past couple of decades in order to curb the ever-growing pharmaceutical expenditure and its consequence on the budgets of public payers, the spending in the out-patient sector has grown by 76% on average from 2000 to 2009 (and went from approximately €260 to €340 in purchasing power standard per capita).[3]

With the advent of the economic crisis of 2008, health-care expenditures became a major target of budget cuts and austerity measures. Price cuts, alterations of the copayment ratio, accession of the value-added tax as well as emendation of the distribution margins were among the tools put to use in order to bridle the public spending on medicines.[3] In 2010 through 2011, 89 initiatives of the kind were implemented in 23 European countries, most notably in Iceland, the Baltic states (Estonia, Latvia, and Lithuania), Greece, Spain, and Portugal.[3]

Among these, external reference pricing (ERP) (also referred to as "external price referencing," "international price benchmark," "external price benchmark," "external price linkage," or "international price linkage") has rapidly become a widespread cost containment tool put to use by the EU MS for their pricing purposes,[3,4] as well as by other countries such as Brazil, Jordan, South Africa, Japan, Turkey, Canada, and Australia that refer to the EU MS drug prices in order to establish their own.[2,5]

The WHO Collaborating Centre for Pricing and Reimbursement Policies defines ERP as "the practice of using the price(s) of a medicine in one or several countries in order to derive a benchmark or reference price for the purposes of setting or negotiating the price of the product in a given country."[6] Consequently, the change of price for a given product in one country affects the price in other countries.

Ever since the inception of ERP as a cost containment tool, although it has been widely accepted and utilized, major concerns have been raised about the potential ramifications on the affordability and patient access to medicines as well as on the industry's revenue and sustainability.[2,4,5,7–13]

7.2 ERP IN EUROPE

In Europe, almost all countries apply ERP leaving out the United Kingdom and Sweden. The latter had ceased to use ERP in 2002 and introduced value-based pricing (VBP) instead.

Denmark has one of the longest experiences with the use of ERP. They have withdrawn from it in 2005 but then introduced it again in the hospital sector starting from 2009.[14] Most of EU countries systematically resort to ERP for the purpose of setting the price of a new drug. ERP is used as a supportive method in Belgium, Finland, Italy, Poland, Spain, and Germany.

In Italy, ERP is used to obtain additional information during the price negotiation, whereas in the past it had been used as the main tool for the assessment of new drugs.

In Germany, ERP has been adopted since 2011 as one of many criteria used to determine the reimbursement price.

In Belgium, ERP is used to obtain supporting detail for the pricing decision. However, price cuts have been introduced in 2013 on the international prices (Austria, Finland, France, Germany, Ireland, and the Netherlands) for reimbursed and patented drugs, authorized on the market for less than 5 years.

In Spain, ERP is used to determine the price of medicine for which there is no alternative available on the Spanish market.

7.2.1 NATIONAL LEGAL FRAMEWORK

The use of ERP whether as a chief tool or as a supportive method to determine the price for new drugs entering the market in a vast majority of countries is supervised by health authorities within a preset legal framework.

In France, ERP methods are agreed within a framework agreement between the Healthcare Products Pricing Committee and the pharmaceutical companies. A similar framework agreement has been cosigned in Ireland between the Irish Pharmaceutical Healthcare Association Ltd. and the Department of Health and Healthcare Executives.

In Spain, since the promulgation of the Decree Law 16/2012 in 2012, ERP is no longer supported by any legislation. However, it still conforms to the internal criteria of the Inter-Ministerial Pricing Committee.

Altogether, ERP methods and rulings are outlined with contrasting levels of accuracy within the national pricing regulations depending on the countries and the level of priority devoted to this tool. Portugal and Austria[15] are good examples of countries in which the legislation provides ample details on the use of ERP. German and Estonian laws provide much less guidance on the matter.

7.2.2 Scope of ERP

On one extreme, Luxembourg resorts to ERP to determine the price of all new marketed drugs. However, ERP is more often applied only to specific groups of medicines such as those publicly reimbursed, prescription-only medicines, or innovative products. As a matter of fact, many countries (Austria, Croatia, Czech Republic, Finland, Ireland, Italy, Latvia, Lithuania, Malta, Poland, Slovakia, Slovenia, and Switzerland) apply ERP solely for publicly reimbursed medicines. Estonia, France, and Germany resort to ERP in the case of innovative and publicly reimbursed medicines only.

Once more, the different legislations describe with disparate levels of thoroughness and accuracy the scope of products subject to ERP. Although it is clearly stated in Danish law that ERP can be used singularly to determine the price for hospital-only medicines, many other countries do not specify whether ERP should be used in the out-patient sector alone as well.

Furthermore, the rules and regulations do not always specify whether or not ERP should be applied to off-patent drugs. Belgium, Cyprus, Estonia, Finland, France, Germany, Greece, Hungary, Norway, and Portugal use ERP in the case of in-patent, or innovative, medicines alone, whereas Austria, Croatia, Iceland, Italy, Slovenia, and the Netherlands use ERP to assess the monetary value of both in- and off-patent medicines.

The European Generic Medicines Association (EGA) refers to the use of ERP on off-patent drugs as uncommon and currently restricted to Bulgaria, Czech Republic, Slovakia, Slovenia, Latvia, Lithuania, Poland, Romania, and Croatia.

7.2.3 Composition of the Country Basket

In ERP, the group of foreign countries looked at in order to determine the price for a new drug is referred to as a country basket. Traditionally, a country basket is appointed with regard to economic comparability or geographic proximity. The number of enlisted countries varies extensively from one country to another. As it happens, Luxembourg has a single country of reference on its list. Croatia, Estonia, Portugal, and Slovenia each have a three-unit country basket, whereas this number mounts up to 31 in the case of Hungary and Poland.

In addition, although a majority of MS of the EU chooses to enlist only EU countries in their basket, a few states such as Hungary, Denmark, Poland, and Finland include the MS of the European Economic Area. Hungary and Poland also index their prices on Switzerland's.

Luxembourg and Estonia refer to the medicine's price in the country of origin. Cyprus, Lithuania, and Romania also do the same when a reference price is not available in one of the countries enlisted in their wonted country basket. However, it is not clearly stated whether that means the country of origin of the drug manufacturer or of the marketing authorization holder in Europe.

Belgium uses ERP as a supportive method and is reported to survey either the price in the country of origin or the average price of its own enlisted countries of reference (26 EU MS).

Overall, France is predominantly cited as a reference by as many as 19 countries in Europe. The United Kingdom and Germany are the second and are quoted by up to 17 countries. The least referenced countries in Europe are Croatia which entered the EU in July 2013 (five countries) and non-MS of the EU countries such as Switzerland (two countries), Iceland (three countries), and Norway (six countries).

7.2.4 Price Calculation and Selection of Reference Products

As mentioned previously, the ERP precepts and regulations are outlined in each country's legislation with differing levels of detail. For instance, the rules concerned with selecting a product of reference are not always set forth intelligibly with regard to variables such as pharmaceutical forms, dosages, pack sizes, out-patient/hospital-only drugs, generics/patent drugs, and reimbursed/nonreimbursed drugs.

Furthermore, the different methods used to calculate the reference price for a drug are not always disclosed comprehensibly (Germany and Estonia) and may differ from one type of product to another within the same country (Croatia and Iceland).

A typical method used (by Austria, Belgium, Cyprus, Denmark, Iceland, Ireland, Portugal, Switzerland, and the Netherlands) to compute the reference price for a new drug is to calculate the average price in a basket of reference countries. Another one is to select the lowest selling price among a list of reference countries (used by Bulgaria, Hungary, Italy, Romania, Slovenia [for original drugs and biosimilars], and Spain). And yet another one is to pick out three or four of the lowest selling prices among a list of reference countries and calculate their average. (Greece, Norway, and Slovakia resort to this method along with the Czech Republic [only to calculate the maximum price].)

France has a list of only four countries of reference and overall applies adjoining rates.

Malta applies double standards. Although the average wholesale price of the country basket is used in the public sector, a specific algorithm is used to calculate the price for the private sector.

In case the price is not available in one or more of the reference countries or not approved by all of them, a few MS (Bulgaria, Croatia, and Cyprus) use the pricing methods of reference countries as alternatives. Other MS (Belgium, Denmark, and Latvia) calculate the price based on the available data while stipulating that the decision will be reconsidered if and when a new reference becomes available in one of the enlisted countries. In the Netherlands, a medicine will be considered for pricing purposes only if a comparable product is available in at least two of the four reference countries. Romania has 12 countries of reference in its basket, and if no data is available in these countries, the price in the country of origin will be considered as an alternative.

Moreover, the prices are subject to frequent updates. In 2012, Ireland hauled down its price levels to match the currency-adjusted average of the ex-factory prices on its list of countries of reference, in compliance with the Supply Terms Conditions and Prices of Medicines' section of the framework agreement between The Irish Pharmaceutical Healthcare Association Ltd. and the Department of Health and Healthcare Executives.

The Norwegian Medicines Agency yearly updates the maximum price of the 250 active ingredients with the highest turnover clinching to reflect the change in European prices with accuracy. In Slovenia, price updates may occur up to twice a year.

The EU MS are predominantly concerned with ex-factory price levels and use public official price databases to gather information. However, pharmacy purchasing price and pharmacy retail price can be considered as well.

As for the package size and the dosage, because they may vary significantly for the same product, the price is often established by referring to a product with the same or the closest dosage and pack size. This practice often generates rather approximate measures and raises a concern in terms of representativeness. Basing the price comparisons on identical pack size would imply the exclusion of some reference countries, but also to ignore the representativeness of the matching pack size for the price level in the reference country.

Many countries do not consider the variations in the pharmaceutical formulation of a drug for ERP purposes (e.g., Latvia, Portugal, and Slovakia), whereas other countries (Belgium, Hungary, and Iceland) will only consider a reference product that has a same or a similar formulation. In these cases, solid forms such as capsules and tablets can be used as reference to determine the price for other solid forms, but cannot be considered to assess other forms of medicine such as injectable.

Many MS (e.g., Austria, Belgium, and Portugal) use the reference price of nonreimbursed products for ERP purposes. Along those lines, and as a rule, the branded version of a product will be selected as reference even though a generic is available in one or many of the reference countries.

7.3 ERP PROCESSES IN NON-EUROPEAN COUNTRIES

Countries from outside Europe such as Australia, Canada, Japan, South Korea, Mexico, New Zealand, and Turkey among other MS of the Organization for Economic Co-operation and Development (OECD) resort to ERP as well, and enlist many EU MS on their respective country baskets. The total amount spent on pharmaceuticals in these countries equals about 80% of the EU pharmaceutical expenditure.

The prices of drugs are not controlled in the United States (US). On that account, there is no need for ERP. However, the US appears on many European and non-European countries' list of references.[7] Furthermore, the aggregate spending of the US and the aforementioned countries mounts up to twice the European pharmaceutical expenditure (i.e., the 25 MS of the EU, Switzerland, Norway, and Ireland combined).

7.3.1 AUSTRALIA

The Pharmaceutical Benefits Pricing Authority (Pricing Authority) determines the selling price for the pharmaceuticals listed under the Pharmaceutical Benefits Scheme.

It uses ERP among several pricing methods and considers the United Kingdom and New Zealand as countries of reference.[7,16]

7.3.2 CANADA

The Patented Medicine Prices Review Board (PMPRB) supervises the ex-factory prices for prescription and nonprescription patented drugs marketed in Canada.

In 1987, the PMPRB endorsed ERP as a chief pricing method for innovative drugs[7] (categorized as a breakthrough, a significant improvement, or a moderate improvement). The Canadian reference country basket was established with consideration to economic and geographic similarities as well as a common target of promoting research and innovation in the pharmaceutical sector.[7,28] It includes the US, France, Germany, Italy, Sweden, Switzerland, and the United Kingdom.

The "maximum average potential price"[17,18,28] for a new patented drug is determined by calculating the median of the ex-factory prices for the same patented drug with the same strength and dosage in each of the seven listed countries.

If the product is available in less than five enlisted countries, the median international price will be considered on an interim basis and reassessed after 3 years.

If the drug is available in an even number of enlisted countries, the median is determined by calculating the average of the two middle prices.[28] If the drug is not available in any of these countries, the most similar strengths of comparable dosage forms of the same patented drug will then be considered.[30]

7.3.3 JAPAN

Japan looks to ERP as a price adjustment tool when the local price diverges significantly from the concurrent price levels in France, Germany, the United Kingdom, and the US.

When the price of a new patented drug on the Japanese market with no replacement drug available, or with a significant added value over all comparable products drops under 75% of the existing price in these countries, the product is systematically reassessed and the price is inflated.[5,7,10,19] Similarly, the price levels drop when they are found to be 1.5 times greater than the average concurrent price overseas.

7.3.4 SOUTH KOREA

South Korea referred to the average ex-factory price in seven countries (the US, the United Kingdom, France, Japan, Germany, Italy, and Switzerland) to negotiate the price of patented drugs up to 2006. Then, in order to curb the ever-growing health-care spending, the government introduced "The Drug Expenditure Rationalization Plan." Currently, price negotiations occur between the National Health Insurance Corporation and the pharmaceutical manufacturers based on price–volume considerations.[*] Individual prices in referenced countries (Australia, France, Germany, Italy,

---

[*] Creativ-Ceutical internal proprietary database.

Japan, Singapore, Spain, Switzerland, Taiwan, and the United Kingdom) are now used in pricing negotiations instead of a formula-based ERP.[20,21]

7.3.5 Mexico

Mexico derives its reference price from the average ex-factory price in the countries where the medicine has sold most during the previous quarter. An external auditor is commissioned to determine the average price for sales to the public applying the following ratio: PRVP (*Precio De Referencia Para Venta al Publico*) = ERP × 1.72.[5,7]

This ratio reflects the combined average margins for wholesale and retail. Nevertheless, Mexico is reported to use the UK prices as a reference.

7.3.6 New Zealand

ERP appears to be conducted in an informal fashion in New Zealand. The country basket includes Canada, Austria, and the United Kingdom.[7]

7.3.7 Turkey

In 2004, over concerns about the growing health-care expenditure, the IEGM (General Directorate for Pharmaceuticals and Pharmacy) became solely responsible for medicine pricing and introduced ERP in Turkey. The maximum ex-factory price in Turkey cannot exceed the lowest price for the identical product in one of the five countries (France, Greece, Italy, Portugal, and Spain).

The enlisted countries of reference have been selected with regard to the characteristics of both their pharmaceutical market and their population in terms of age range distribution and health state.

If the product is not available in any of these countries, another EU MS is used as an alternative reference. The price in the country of origin is considered as a reference price if the product is not marketed in the EU, or if it is lower than the ex-factory price in all the enlisted reference countries.

If the product is exclusively marketed in Turkey, negotiations are conducted to determine the price.

The price for all generics and drugs whose patents expired is set at 66% of the cheapest originator product available in any of the six countries of reference.[22,23]

7.4 CONCERNS RELATED TO ERP

Throughout the years, the ERP principles and methods of application have raised a good deal of concerns and castigations among the different stakeholders of the pharmaceutical market.

Many have denounced it for creating "path dependence" as it operates like a self-contained and market blind system. In fact, the price levels on the pharmaceutical market are determined by the system itself through mutual observation between countries, leaving out such factors as the disparities of health needs, as well as income and health-care costs between these countries.[8,9]

Also, the ease of access to the relevant data sources and the accuracy of price information are the major drawbacks to ERP.[11]

The available price-level indicators may vary from one country to another. For instance, the United Kingdom and the Netherlands only provide information on the pharmacy purchasing price. Because the ex-factory price is the one used by most countries for reference pricing purposes, they must derive it from the available data by calculating and subtracting the wholesale margin, which implies a certain level of conjecture and a margin of error.[5,8,10,11,24]

Along those lines, some of the publicly available pricing data is essentially inaccurate as it does not account for the managed entry agreements and the confidential yet quite substantial discounts conceded by the manufacturers.[5,8,10,11] This disputable practice results in inflated rates mostly in the case of drugs that enjoy monopolistic position.

Furthermore, due to untidy monitoring, the price cuts conducted by one or several countries selected as a reference do not systematically reverberate in other countries that should be concerned.

It is reported in the literature that the same product can be launched in different countries with different commercial designations, different proportions, and different dosages in an attempt by the manufacturer to cramp the foreign health authorities in their research for a reference product.[5,7,10,37] In addition, prices are often listed in local currencies and subject to the volatility of the exchange rates.[4,8] During the past 5 years, drug prices have decreased dramatically in Switzerland, boosted by the depreciation of the respective currencies of the Swiss reference countries.[25]

Finally, countries that refer to non-Euro-zone do not divulge the employed exchange rates, which results in price miscalculations in third-party countries.

7.4.1 Potential Consequences of ERP

The influence of ERP on the pharmaceutical market is quite uncertain and open to debate.[11,27] The industry is overall concerned with the spillover effects of one country on another, and they often argue that ERP results in a downward price convergence.

Indeed, as a consequence of the massive use of ERP, a low entry price conceded by a manufacturer in one territory can constrain him or her to reconsider his or her pricing policies everywhere.[4,5,8,11,37] Two recent studies however suggested that there is no substantial decline in the general price imbalance between EU MS.[26,27]

The first study looked at over a thousand prescription drugs from 36 therapeutic categories in 30 countries (EU and non-EU countries) over 12 years. It revealed that half of the price differentials exceeded 50% over time and that parallel trade did not affect the price distribution among the EU MS.

The second one focused on 10 patent drugs in 15 European countries over 5 years. It pointed out a divergence in pricing between 2008 and 2012, driven by Germany and Greece. However, it ended up supporting the idea of a price convergence nuanced by a substantial difference between the lowest and the highest price countries.

Both studies combined show that even though ERP arguably results in a price convergence over time, price differences still prevail under the effect of various ERP methodologies as well as the use of other pricing methods and policies.

7.4.1.1 Patient Access to Medicine

ERP is often deprecated for its potential impact on patient's access to medicine. It represents an incentive for manufacturers to implement a "launch sequence strategy" in order to postpone or avoid launching a new drug in smaller and lower priced markets that might be referenced by other countries.[4,5,8,10]

For instance, the pharmaceutical companies systematically delay the entry of their products to the Belgian market, which is considered to be in the low EU range and has a considerable impact through ERP on other more profitable countries.[28]

Moreover, the extensive application of ERP is reported to have a circular pricing effect. Because more and more countries are cited as reference, it becomes harder to designate which country the price originates from. On top of that, price alterations, in one country, may, at least theoretically, trigger a domino effect in other countries. Altogether this puts more incentive on the strategic launch.[37]

It is difficult to assess the impact of launch sequence strategies on patient access to medicine in low-end markets, as they are in fact less attractive to suppliers. However, the delays they must cope with might as well be the result of a parallel trade between countries. The two factors certainly coexist and their respective footprints are hard to disjoin.[5,8]

7.4.1.2 Affordability

Countries with the highest absolute price levels (Germany, Denmark, Ireland, and Italy) reportedly have low relative price levels considering their respective gross domestic product (GDP) per capita, whereas countries such as Poland, Romania, and Bulgaria have low prices in the absolute value, yet they pay relatively much with regard to their GDP.

Price convergence is one of the factors contributing to this occurrence, as pharmaceutical companies seek to obtain convergent entry prices in all markets with no consideration for the local purchasing power in order to contain or eliminate the parallel trade of medicines between countries.[29]

Countries with the lowest price levels often suffer from stock shortages caused by a massive parallel outgoing trade. In 2012, 200 products (strengths, pack sizes, and chemical entities) were lacking from the market in Bulgaria.

ERP methods primarily aspire to achieve a price control and a rapid devaluation of a medicinal product. However, these methods combined with the parallel trade of medicines between countries become counterproductive as they impel the pharmaceutical companies to set a high target price for the sake of preserving their revenues and sustainability.[4,5,37]

7.4.1.3 Industry Revenue and Sustainability

The Ramsey pricing principles suggest that the prices should vary across the different markets according to the demand elasticity, and the users should be charged with regard to their price-sensitiveness.[11] This is reported to be an efficient method to provide for the global joint costs of pharmaceutical research and development (R&D) as it allows for the pharmaceutical companies to achieve large volume sales in poorer countries without jeopardizing their revenues in the richer and less price-sensitive countries.[29]

However, applying these principles would require disavowing the VBP methods based on effectiveness and/or cost-effectiveness as well as ERP, which is supposed to discourage incremental innovation[12,13,29] by decreasing revenues and leaving the industry with lesser resources to invest in R&D and by limiting these activities to niche markets. It is unclear yet to what extent ERP impacts the activities of research and development. And it seems that no measure of this impact can be taken in the near future.[13] Nevertheless, it appears that ERP has a massive and negative impact on the overall competitiveness of both the in-patent and off-patent branches of the industry.

The EGA upholds that ERP limits its industry's potential to approach specific markets as it takes the prices down to unsustainable levels. As a case in point, the EGA evokes the Olanzapine price in Bulgaria which dropped by 98% after applying ERP and referring to the Danish price, which in turn caused the manufacturers to withdraw from the Bulgarian market and consequently restricted the Bulgarian patients' access to this medicine. The EGA goes on to highlight that referring to the price levels in countries that foster a procurement and tendering system that takes down the prices to unsustainable levels is detrimental to the generics' branch of the industry limiting both the patient's access to the medicine and the potential savings made by the payers as they purchase generic drugs instead of patented ones.

However, the European Federation of Pharmaceutical Industries and Associations (EFPIA) is concerned with the indirect and adverse effects of ERP, particularly in the context of budget crisis and proliferation of cost containment measures.

Charles River Associates (Boston, MA) conducted a pair of studies that illustrate this point of view.

The first one attempted to assess the impact of ERP and parallel trade on social welfare and patient access to medicine. It came to the conclusion that ERP and parallel trade created a reverse spillover effect from the low-end markets up to the top, bringing the average prices down, with overall limited benefits to the payers in terms of cost savings, but with a considerable impact on patient access in the lowest priced markets. Furthermore, the study concluded that ERP and parallel trade had a noticeable impact on the industry's potential to invest in R&D.

Again, to illustrate the spillover effect of ERP, the EFPIA put forward the example of Greece. They calculated that if all countries citing the Greek price levels as reference had updated their own prices following the 10% price drop of 2011, the overall financial impact on the industry would mount up to €299 million in Greece, €799 million in Europe, and €2,154 billion worldwide.

The second study looked at the Swiss pricing regulations and the international impact of a price cut in Switzerland. According to this study, the spillover effect of a 10% average price cut in Switzerland would have a financial impact estimated at €430 million in Switzerland and 495.2 million worldwide. It also proceeded to modeling the impacts of the price cuts in Switzerland on patient access in Canada. It came to the conclusion that launching a new drug in Switzerland first resulted in raising the Canadian launch price. On the contrary, delaying the launch of the new drug on the Swiss market granted better access to the Canadian patients.

Furthermore, the EFPIA put forward the example of price miscalculations that occurred recently in Greece. The consequences of this anomaly reverberated in 13 different countries that quoted the Greek price as reference.

7.5 VBP AND ERP

VBP is another common pricing tool among European countries, and as such, it is interesting to measure the differences between it and ERP.

VBP can be defined as a model in which the price is set on the perceived value to the consumer. In that understanding, the notion of value depends on the consumer's appreciation, and the selling price is determined by coupling the delivered value and the perceived one.

Public payers entrust health technology assessment (HTA) organizations to establish evidentiary assessments of the added value of a new medicinal product by comparison with a reference therapy.

Although the core notion of value is essentially the same, the perceived value of a product differs from one country to another depending on the particularities of the health-care service, the medical practice, the availability of different comparable products, and so on. Furthermore, the set of evidence required to assess the value of a new medicine may vary sensibly from one country to another because of the different methods of assessment and the criteria considered in determining the benefits, the choice of comparators, and so on.

In France, the Transparency Committee establishes the Improvement of Actual Benefit (*Amélioration du Service Médical Rendu* [ASMR]) by assessing the level of clinical improvement of the new medicine as opposed to the existing treatment and with a focus on measuring the improvement in absolute effect size. In Germany, the rate ratio of relevant endpoints is measured instead, and the upper 95% confidence interval is considered to acknowledge the drug value but not the estimate of the rate ratio. In the United Kingdom, the National Institute for Health and Care Excellence uses the cost per quality-adjusted life year to assess the health benefits delivered by a new drug and upholds an incremental cost-effectiveness ratio (ICER) threshold range. Many other countries use the ICER among other criteria but without defining an ICER threshold for a new technology approval.

Thus, value is perceived and measured differently from one country to another.

Furthermore, VBP can be calculated *ex-ante*, which means before price setting (France), or *ex-post*, which means after price setting (Sweden and the United Kingdom). The use of VBP entails that a higher value drug will enjoy a higher price and implies the use of national reference pricing. However, excluding the countries that have a legal threshold associated with the use of ICER, it is not clearly defined how the added

value of a product is acquainted with a willingness to pay a higher price.

7.5.1 ERP AS AN ALTERNATIVE TO VBP

ERP is a relevant alternative to VBP particularly in the case of a country that wishes to use benchmark pricing methods, yet it does not have sufficient resources to conduct VBP, has a better use for these resources, or does not consider VBP as an efficient pricing tool.

7.5.2 ERP AND VBP COMBINATION

Many countries use ERP to complement VBP methods as it is hard to establish a systemic correlation between the assessed added value of a product and the willingness to pay. Furthermore, health authorities have not always settled for ERP on their own initiative. As a matter of fact, the manufacturer union in France influenced the health authorities' decision to introduce ERP in order to guarantee that the French prices remain within the EU range and prevent the spillover effects on other countries.

In general, both VBP and ERP methods are open to debate. Because the perception of value differs from one country to another, modulating a value based on the willingness to pay of another country is in fact a disputable proposition. It may seem absurd for instance to set a price for a new product based on the perceived value and then go on to amend that rate with reference to price levels in other countries.

However, assessing the value of a new product and setting the right price for it is a complex and delicate endeavor. As such, in order to make an informed decision, it may not be unreasonable to monitor the price levels in other countries to see if the product is not dramatically under- or overestimated. To that end, ERP can be put to use efficiently to modulate the VBP decisions. However, whenever ERP is used as a chief assessment tool, it cannot be complemented with VBP methods because the price is determined by benchmarking methods in the first place, regardless of the perceived value to the consumer.

ERP and VBP can also coexist while having distinct prerogatives. In some instances, ERP applies to innovative products alone, whereas all other products are assessed with VBP methods. Under this configuration, the HTA organizations are again entrusted to tell apart the innovative from the noninnovative drugs.

REFERENCES

1. Council Directive of 21 December 1988 relating to the transparency of measures regulating the pricing of medicinal products for human use and their inclusion in the scope of national health insurance systems (89/105/EEC). (This directive is under review to adapt it to the current pharmaceutical environment). Available from: http://eurlex.europa.eu/LexUriServ/LexUriServ.do?uri=CELEX:31989L0105:en:HTML (Cited October 26, 2016).
2. Leopold C, Vogler S, Mantel-Teeuwisse AK, de Joncheere K, Leufkens HG, Laing R. (2012). Differences in external price referencing in Europe-A descriptive overview. *Health Policy* 104:50–60.

3. Vogler S, Zimmermann N, Leopold C, de Joncheere K. (2011). Pharmaceutical policies in European countries in response to the global financial crisis. *Southern Med Review* 4(2):22–32. Available from: http://apps.who.int/medicinedocs/documents/s19046en/s19046en.pdf (Cited August 14, 2013).

4. European parliament. Differences in costs of and access to pharmaceutical products in the EU (2011). Available from: http://www.europarl.europa.eu/document/activities/cont/201201/20120130ATT36575/20120130ATT36575EN.pdf (Cited August 14, 2013).

5. Espin J, Rovira J, De Labry AO. WHO/HAI project on medicine prices and availability-Working paper 1: External reference pricing (2011 May). Available from: http://www.haiweb.org/medicineprices/24072012/ERPfinalMay2011.pdf (Cited August 15, 2013).

6. The WHO Collaborating Centre for Pricing and Reimbursement Policies, Glossary. Available from: http://whocc.goeg.at/Glossary/PreferredTerms/External%20price%20referencing (Cited August 14, 2013).

7. OECD. Pharmaceutical Pricing Policies in a Global Market. (2008 Sep). Available from: http://www.oecd.org/fr/els/systemes-sante/pharmaceuticalpricingpoliciesinaglobalmarket.htm (Cited August 14, 2013).

8. Kanavos P, Espin J, van der Aardweg S. (2010 Jan). Short- and long-term effects of value-based pricing vs. external price referencing. *EMINET*. Available from: http://ec.europa.eu/enterprise/sectors/healthcare/files/docs/valuebased_pharma-pricing_012010_en.pdf (Cited August 14, 2013).

9. Cueni TB. International Price Referencing-Is there a "right" way to perform it? Presentation. *ISPOR 15th Annual European Congress*, Berlin, Germany (November 3–7, 2012).

10. OECD. Improving health-system efficiency: Achieving better value for money-Ensuring efficiency in pharmaceutical expenditures: policies to improve value for money. Joint OECD/European Commission conference, Brussels (September 17, 2008). Available from: http://ec.europa.eu/social/main.jsp?catId=88&eventsId=106&langId=en&moreDocuments=yes&tableName=event (Cited August 14, 2013).

11. Bouvy J, Vogler S. Pricing and Reimbursement Policies: Impacts on Innovation. Background Paper 8.3 (2013 May 23). Available from: http://www.who.int/medicines/areas/priority_medicines/BP8_3_pricing.pdf (Cited 2013 Sep 12).

12. Eucope (European Confederation of Pharmaceutical Entrepreneurs). Explanatory memorandum. Pharmaceutical prices: Why are there differences between Member States (2012). Available from: http://www.eucope.org/en/files/2012/10/EUCOPE-IRP.pdf (Cited August 30, 2013).

13. Brandt L. Price tagging the priceless: International reference pricing for medicines in theory and practice. ECIPE Policy Briefs N° 03/2013. Available from: http://www.ecipe.org/media/publication_pdfs/ECIPE_Policy_Brief_IRP_30_May_FINAL_pdf.pdf (Cited September 12, 2013).

14. Kaiser U, Méndez SJ, Rønde T, Ullrich H. Regulation of Pharmaceutical Prices: Evidence from a Reference Price Reform in Denmark. IZA DP No. 7248. Discussion Paper (2013 Feb). Available from: http://ftp.iza.org/dp7248.pdf (Cited September 12, 2013).

15. Kaiser U, Méndez SJ, Rønde T, Ullrich H. Regulation of Pharmaceutical Prices: Evidence from a Reference Price Reform in Denmark. IZA DP No. 7248. Discussion Paper (2013 Feb). Available from: http://ftp.iza.org/dp7248.pdf (Cited September 12, 2013).

16. Australian Government. Department of Health and Ageing. PBPA Policies, Procedures and Methods. Available from: http://www.health.gov.au/internet/publications/publishing.nsf/Content/pbs-pbpa-policies-contents~pbs-pbpa-policies-preamble (Cited August 30, 2013).

17. Prashant Y. Differential Pricing for Pharmaceuticals Review of current knowledge, new findings and ideas for action. A study conducted for the U.K. Department for International Development (DFID) (2010 Aug). Available from: https://www.gov.uk/government/uploads/system/uploads/attachment_data/file/67672/diff-pcing-pharma.pdf (Cited August 30, 2013).

18. Richards C. Canada: Healthcare System and Drug Regulatory Overview. DataMonitor. DMKC0060758 (2012 Mar 15).

19. Inazumi Y (Chief of drug price, Economic Affairs Division, Health policy Bureau, Ministry of Health, Labour and Welfare, Japan). Drug evaluation and pricing. Presentation (2008).

20. Jones RS. Health-Care Reform in Korea. OECD. Economics Department Working Papers, No. 797, OECD Publishing (2010). Available from: http://dx.doi.org/10.1787/5kmbhk53x7nt-en (Cited August 30, 2013).

21. Kwon S. Health Care Reform in Korea: Key Challenges. IMF Conference (2011 Oct 3). Available from: http://www.imf.org/external/np/seminars/eng/2011/healthcare/pdfs/s3_kwon.pdf (Cited August 30, 2013).

22. Richards C. Turkey: Healthcare System and Drug Regulatory Overview. DataMonitor. DMKC0060716 (2012 May 15).

23. Koçkaya G. Pharmaceutical Policies and Market Access in Turkey. ISPOR Connections (2012). Available from: http://www.ispor.org/news/articles/april12/pharmaceutical-policies-market-access-turkey.asp (Cited August 30, 2013).

24. Carone G, Schwierz C, Xavier A. Cost-containment policies in public pharmaceutical spending in the EU. European Commission-European Economy. Economic papers 461 (2012 Sep). Available from: http://ec.europa.eu/economy_finance/publications/economic_paper/2012/pdf/ecp_461_en.pdf (Cited August 30, 2013).

25. The Pharmaletter. Swiss pharma market shrank for first time in 2010 (2011 Feb). Available from: http://www.thepharmaletter.com/file/101871/swiss-pharma-market-shrank-for-first-time-in-2010.html (Cited September 12, 2013).

26. Kyle MK, Allsbrook JS, Schulman KA. 2008. Does Reimportation Reduce Price Differences for Prescription Drugs? Lessons from the European Union. *Health Services Research* 43(4):1308–1324.

27. Leopold C, Mantel-Teeuwisse AK, Vogler S, de Joncheere K, Laing R. Leufkens HG. Is Europe still heading to a common price level for on-patent medicines? An exploratory study among 15 Western European countries. *Health Policy* (2013 Sep).

28. Maervoet J, Toumi M. Time to Market Access for innovative drugs in England, Wales, France and Belgium. *ISPOR 15th Annual European Congress*, Berlin, Germany (2012 Nov 3–7).

29. Garau M, Towse A, Danzon P. Pharmaceutical pricing in Europe: Is differential pricing a win-win solution? Office of health economics. Occasional paper 11/01 (2011 Feb 11). Available from: http://www.ohe.org/publications/article/pharmaceutical-pricing-in-europe-10.cfm (Cited September 12, 2013).

8 Gap between Payers and Regulators

8.1 INTRODUCTION

Medicine regulators are entities that typically do not consider the cost of treatments, but rather focus on internal validity of data from clinical trials such as efficacy, safety, and the pharmaceutical quality. For the purpose of this chapter, we will not consider the quality that has to be granted to gain a marketing authorization (MAu). In contrast, payers are generally entities that finance or reimburse the cost of health products and services. They tend to focus on external validity of clinical trial data, comparative effectiveness, real-life treatment pathways, and cost-effectiveness. These two authorities have often different perspectives on uncertainty and risk.

In order to gain market access (MA), a new drug should satisfy the requirements of both regulators and payers. Although regulators are in charge of assessing the drug's safety and clinical efficacy, health technology assessment (HTA) bodies and payers are more concerned with the therapeutic added value, cost-effectiveness, and budget impact of the drug. To make their decision on the entry of a new drug to the market, regulators assess the drug on its own merit and need objective data on safety, efficacy, and quality without taking into account the economic considerations. The regulator will assess the benefit:risk ratio while considering clinical trials with high internal validity and free of bias, whereas payers need information on health and cost consequences associated with the drug once used on the market in comparison with one or more reference treatments expected to be the next best alternative(s). Clinical trials with high internal validity inform efficacy (measured in an experimental context), whereas payers are interested in efficacy under the normal condition of use also called effectiveness. The transition from efficacy to effectiveness is illustrated in Figure 8.1. Therefore, in the process of health benefit assessment, HTA bodies/payers often conduct their own analysis driven by effectiveness, instead of relying on regulators' assessment driven by efficacy, which might lead to different conclusions.

Both regulators and payers face the challenge of balancing early MA and a stringent benefit/risk assessment. In this process, they inevitably have to deal with uncertainty about safety and efficacy parameters of the drug. Recent regulatory decisions suggest that regulators accept uncertainty as a part of the decision making and use well-established and standardized management tools to facilitate early access of treatment to patients, whereas payers, who are under budget constraints, have less tolerance to uncertainty and their management strategies are often directed toward cost containment. If the level of uncertainty remains higher than their acceptability threshold, they are likely to refute or limit the reimbursement to a restricted population. These discrepancies in decision-making drivers lead to a gap between licensing

approvals by regulators and reimbursement decisions, which can be delayed until HTA bodies or payers obtain additional data from manufacturers that can minimize uncertainty. Payers manage broader uncertainty and dispose of less robust tools to manage uncertainty.

The objective of this chapter is therefore to compare the management of uncertainty between regulators and payers, and the different tools they use to assess the appropriate balance between the need for rapid access to innovative drugs and the need to ensure an appropriate mitigation of uncertainty.

8.2 UNCERTAINTY VERSUS RISK

Terms "uncertainty" and "risk" are often confused, and it is important to differentiate between these two concepts. Uncertainty is a situation in which the current state of knowledge is such that the order or nature of things is unknown; consequences, extent, or magnitude of circumstances, conditions, or events are unpredictable; and credible probabilities to possible outcomes cannot be assigned.[1] In contrast, risk is characterized by at least possible outcome(s) that are identified and a probability for possible outcome(s) can be assigned. Uncertainty is associated with vacuum and is a strong source of uncomforted perception and fear leading to rejection of product coverage or approval. Risk is less worrying if the events that may happen are identified and the likelihood of their occurrence is reasonably quantified. Then risk is a source of anxiety, and the acceptability of the risk will depend on the nature and probability of occurrence. Assessing the risk enables us to manage it, that is, to prevent events from happening or controlling the extent of the negative consequences of the events. This is called a risk mitigation plan. Because payers can develop a risk mitigation/management plan when they are confronted with a risk rather than uncertainty, they should be approached with information about the former and not the latter.

8.3 PAYERS VERSUS REGULATORS

Regulators have well-established approaches to the assessment and mitigation of risks and uncertainties, in particular for safety issues. However, payers have traditionally accepted that uncertainties and risks exist, but did not feel compelled to develop a systematic approach to postapproval management of various uncertainties and risks. Moreover, whereas regulators deal with well-defined risks that can be transposed across different geographies, payers need to assess heterogeneous risks in their own setting. No structured and universally accepted methods exist that could be employed by all payers.

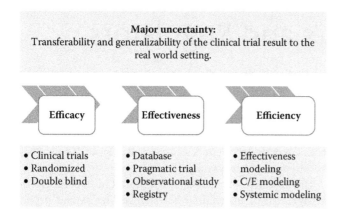

Major uncertainty:
Transferability and generalizability of the clinical trial result to the
real world setting.

| Efficacy | Effectiveness | Efficiency |

- Clinical trials
- Randomized
- Double blind

- Database
- Pragmatic trial
- Observational study
- Registry

- Effectiveness modeling
- C/E modeling
- Systemic modeling

FIGURE 8.1 The transition from efficacy to effectiveness and related kinds of evidence. C/E-cost-effectiveness.

Increasing budgetary constraints and cases of very costly drugs that failed to show effectiveness that would warrant the high price tag have led payers to consider the need to develop new approaches toward such risks. Increased methodological requirements for HTA and cost containment measures are examples of those attempts.

8.4 SOURCES OF UNCERTAINTY

8.4.1 REGULATORS

Uncertainty is inherent to all regulatory approval decisions. This is acknowledged by experts and the society. Therefore, whenever a new drug is approved, it carries some potential uncertainties. In the process of benefit/risk assessment of a new drug, regulators manage uncertainty and risk at all stages of the regulatory review. In the preapproval phase, risks are related to the trial sample size, duration of follow-up, statistical analysis, observed safety signal, package insert scrutiny, and so on. In the postapproval phase, unexpected issues may arise from large postmarketing observational cohorts, database analysis, spontaneous reporting, and misuse.

8.4.2 HTA BODIES/PAYERS

For HTA bodies/payers, most of the uncertainty lies in the transposability of clinical results from one jurisdiction to another, from one setting to another, and from the generalizability of clinical trial result to the broader population that will to receive the drug in the real-world setting. How efficacy in randomized controlled trials (RCTs) translates into effectiveness, and furthermore into efficiency, and budget impact are the issues the payers have to deal with. This can be related to limitations in the trial for the patient population, comparator drug, study design, and health outcomes. It may also come from other sources, such as prescribers' medical practice and patients' behavior such as drug misuse, off-label use, concomitant medication, concomitant disease, patient adherence, and mismanagement. There is also uncertainty related to the context such as the target size population that may be larger than expected leading to wider

budget impact, guidelines evolution, new serotype emergence, prescription transfer, and new competitive environment. These will be discussed in Section 8.5.

8.5 RISK MANAGEMENT OF DRUG VALUE UNCERTAINTY—HTA/PAYER LEVEL

The major uncertainty payers tend to focus on is related to the transferability and generalizability of the clinical trial results (efficacy) into the real-world setting (effectiveness). This can be related to the patient population, comparator drug, and study design or health outcomes.

8.5.1 POPULATION

Clinical trial population differs from the target population in terms of the following:

- Geographic area
- Inclusion/exclusion criteria (e.g., age, severity)
- Compliance

Table 8.1 contains the examples of such uncertainties identified by various HTA bodies.

These issues can be avoided if the company establishes an early dialog with the payers and assesses the patient path in the key countries, compares treatment guidelines, and undertakes population segmentation, so that they can be incorporated in the trial design. The company should also consider setting up a registry of patients nonincluded in trials and also follow-up patients who have dropped out from the trials, as this may be very informative for payers (e.g., What are their characteristics? What was the disease course after they have dropped out? Did they respond to a successive therapy?).

8.5.2 COMPARATOR

Payers are often dissatisfied with the choice of appropriate comparator in the clinical trials. They may criticize the following:

- Difference between guidelines and the real clinical practice
- Off-label use
- Changes in the reference treatment
- Differences in the clinical practice between countries

They are exemplified in Table 8.2.

To prevent those issues, manufacturers should ensure that comparator is approved in the indication, line of treatment, and subgroup population, and that it fits the real-life practice. They should also consider the value of a phase IIIB study to address appropriate comparators for specific HTA agencies when the comparators are not homogeneous between countries. Alternatively, they should be prepared to produce a meta-regression or mixed treatment comparison meta-analysis or network meta-analysis.

TABLE 8.1
Examples of Population-Related Uncertainties Identified by Various HTA Bodies

Drug and Indication	HTA Body Date	Comments
Ruxolitinib Myelofibrosis	HAS January 9, 2013	Transferability of clinical trial results to the current clinical practice is limited due to exclusion of patients who are candidates for hematopoietic stem cell transplantation in the trial COMFORT II.
Ivacaftor Cystic fibrosis	G-BA February 7, 2013	Study submitted does not include severe patients.
Pivmecillinam Urinary tract infections	HAS April 3, 2013	Study results performed in Nordic countries cannot be transposed to clinical practice in France, due to current resistance of *Escherichia coli* to pivmecillinam in France, which is the main germ encountered in community-acquired urinary infections.
Vinflunine Transitional cell carcinoma of the urothelial tract	NICE January 2013	Uncertainty regarding the effectiveness for the whole licensed population in the United Kingdom due to differences in the characteristics and treatment pathways in the trial versus UK clinical practice: The study population was younger and fitter, and had better renal function than the general UK patients with advanced or metastatic transitional cell carcinoma of the urothelial track. The study excluded patients who had had adjuvant or neoadjuvant chemotherapy, even though in UK practice, many patients in the United Kingdom eligible to receive the second-line palliative chemotherapy will already have received two lines of treatment.

Source: Creativ-Ceutical, Paris, Internal Research, 2013.
Note: HAS, Haute Autorité de Santé; NICE, National Institute for Health and Care Excellence; G-BA, Gemeinsame Bundesausschuss.

TABLE 8.2
Examples of Comparator-Related Uncertainties Identified by Various HTA Bodies

Drug and Indication	HTA Body	Comments
Brentuximab vedotin Lymphoma	HAS January 9, 2013	Public health benefit cannot be assessed due to missing data (especially comparative data, but no available alternative treatment) to assess the expected impact on morbi-mortality and quality of life of treated patients
Axitinib Renal cell carcinoma	HAS January 9, 2013	No comparative data versus alternative treatments having MAu after failure to sunitinib
Pivmecillinam Axial spondylarthritis	HAS April 3, 2013	No study compared pivmecillinam to antibiotics currently used in France to treat community-acquired urinary infections (the company submitted one study vs. active comparator [sulfamethizol] and one study vs. placebo)
Linagliptin Type II diabetes	G-BA February 21, 2013	Dual therapy: one potentially relevant direct comparative study, BUT issue in the study design as two different treatment strategies were compared: one strategy without a specific target level for blood glucose for linagliptin versus another strategy with a specific target for blood glucose for glimepiride (glimepiride dose up-titrated in the first phase of the study)

Source: Creativ-Ceutical, Paris, Internal Research, 2013.
Note: HAS, Haute Autorité de Santé; G-BA, Gemeinsame Bundesausschuss.

8.5.3 Design

Trial design issues often raised by payers are as follows: not appropriate study duration, small number of patients, noninferiority versus superiority studies, open label versus blind design, blinding maintenance, and patients switching treatment arm. Those are illustrated in Table 8.3.

To address these issues, manufacturers should ensure that the trial design allows differentiation between treatments, that the study is powered for subgroup analysis and prespecify secondary analysis versus exploratory ones, and that there is double-blinded centralized rating/assessment, use different scale for inclusion and primary outcome, and finally decide on acceptable study duration from the payers' perspective.

8.5.4 Outcome

HTA bodies often regret that trials used nonvalidated instruments, surrogate outcomes, or composite outcomes. These concerns are presented in Table 8.4.

Manufacturers should ensure that outcome instruments are validated or validate them, possibly in parallel along study

TABLE 8.3

Examples of Design-Related Uncertainties Identified by Various HTA Bodies

Drug and Indication	HTA body	Comments
Ruxolitinib Myelofibrosis	HAS January 9, 2013	One open-label trial and doubt on the blind maintenance in the double-blind trial due to adverse events (thrombocytopenia)
Aitinib Renal cell carcinoma	HAS January 9, 2013	Weaknesses in the methodology of the phase III study submitted for the dossier:
		Open-labeled study whereas double-blind study would have been possible
Adalimumab Axial spondylarthritis	HAS February 20, 2013	The transferability of the results to clinical practice is not guaranteed because of the limited duration of the randomized trial (12 weeks)
Aclidinium bromide Chronic obstructive pulmonary disease	G-BA March 21, 2013	Inadequate duration of comparative trials for a long-term therapy (<6 months)
Linagliptin Type II diabetes	G-BA February 21, 2013	It was concluded that this study did not constitute a comparison just of the two drugs, but rather a comparison of two combined interventions (treatment strategy+drug) and that it was not certain that the effects observed in the study were attributable to the respective drugs used (could also solely be due to the different treatment strategies)

Source: Creativ-Ceutical, Paris, Internal Research, 2013.

Note: HAS, Haute Autorité de Santé; G-BA, Gemeinsame Bundesausschuss.

TABLE 8.4

Examples of Outcome-Related Uncertainties Identified by Various HTA Bodies

Drug and Indication	HTA Body	Comments
Ruxolitinib Myelofibrosis	HAS January 9, 2013	Lack of validation of the instrument EORTC QLQ-C30 in this disease
Aclidinium bromide Chronic obstructive pulmonary disease	HAS April 17, 2013	The efficacy of aclidinium bromide on exacerbations and hospitalizations was not evaluated as an endpoint (main or secondary) even though they are considered the important criteria to assess the clinical benefit in the patient.
Ruxolitinib Myelofibrosis	G-BA February 7, 2013	The instrument used to assess the symptoms of myelofibrosis was not enough validated (symptoms were assessed using a symptom diary, MSAF v2.0 [Myelofibrosis Symptom Assessment Form v2.0] defining a total symptom score)

Source: Creativ-Ceutical, Paris, Internal Research, 2013.

Note: HAS, Haute Autorité de Santé; G-BA, Gemeinsame Bundesausschuss.

development, appreciate the predictive value of surrogate endpoints and eventually document it, and avoid composite end points. Composite endpoints should be analyzed toward alternative options: segmented to identify which element is driving the results and validating that element.

8.5.5 INDIRECT COMPARISON

Indirect comparisons are carefully scrutinized by HTA bodies/payers as to the studies selection, homogeneity of studies, consistency of designs and endpoints, methodology used, and model fit, as shown in Table 8.5.

In fact, different HTA bodies have different levels of acceptability of indirect comparison, ranging from high for the United Kingdom and Sweden to low for France and very low for Germany. In any case, the feasibility of the indirect comparison should be anticipated when designing phase II and III trials; the methodology needs to follow HTA-accepted guidelines and include extensive sensitivity analysis.

8.6 RISK MANAGEMENT TOOLS

8.6.1 REGULATORS

To manage these uncertainties, regulators have developed expertise, processes, and tools for risk management adapted from multiple scientific areas (Table 8.6).

The new EU pharmacovigilance legislation allows regulators to require not only postauthorization safety study (PASS) but also postauthorization efficacy study as a condition of MAu.

Regulators may grant conditional approval when the risk–benefit balance is positive, but a part of the clinical data is still awaited. Drugs under conditional approval target seriously debilitating diseases or life-threatening diseases, are used in emergency situations, or are designated as orphan products when the rarity prevents the collection of compelling information. With the new EU legislation, specific obligations may be imposed in relation to the collection of data in terms of nature, objective, and time frame for any study. In some cases, where there is uncertainty regarding transferability of

TABLE 8.5

Examples of Indirect Comparison-Related Uncertainties Identified by Various HTA Bodies

Drug and Indication	HTA Body	Comments
Aitinib Renal cell carcinoma	HAS January 9, 2013	Indirect comparisons provided in the dossier but considered of modest interest
Aclidinium bromide Chronic obstructive pulmonary disease	G-BA March 21, 2013	Indirect comparison data submitted for aclidinium and tiotropium versus placebo were not transparent and unclear: • Some of the manufacturer's information did not match the corresponding original data and could not be traced in the original sources (especially patient numbers and confidence intervals) • Often unclear whether data were issued from the manufacturer or from another source and calculation methods • Out of 24 studies (3 on aclidinium, 21 on tiotropium), only 14 were suitable for the assessment, because the others had lasted for less than 6 months • The manufacturer repeatedly did not consider the study results on relevant outcomes (exacerbations, mortality, quality of life), although available in the original sources
Eltrombopag Immune (idiopathic) thrombocytopenic purpura	NICE June 2013	Considerable uncertainty of the indirect comparison results due to heterogeneity of studies

Source: Creativ-Ceutical, Paris, Internal Research, 2013.

Note: HAS, Haute Autorité de Santé; NICE, National Institute for Health and Care Excellence; G-BA, Gemeinsame Bundesausschuss.

TABLE 8.6

Regulators' Risk Management Tools

Preapproval	Postapproval
• Preclinical toxicological and pharmacological models • Structure activity insight and databases • Biological/immunological profile • Pharmacovigilance • *Ad hoc* studies • Development guidelines	• Conditional approval • Approval under exceptional circumstances • Risk management plans • Spontaneous notifications • Database analysis • *Ad hoc* studies

foreign data, the regulator could decide to issue conditional approval contingent upon the marketing authorization holder (MAH) conducting local community-based trials in less stringently selected "real-world" patients. The purpose would be to evaluate effectiveness (as opposed to efficacy) as well as long-term safety in a practical setting.

In some cases, the drug can be approved, but further studies are requested. The European Medicines Agency (EMA) has adopted Risk Management Plans for all drugs that are granted MAu and has increased the postlaunch proactive demand for pharmacovigilance studies. Risk Management Plans also include risk minimization actions/plans and measuring the effectiveness of the drug.[2]

MAu may be granted under exceptional circumstances when an applicant is unable to provide comprehensive data on the efficacy and safety because of one of the following reasons: The indication is very rare; the present state of scientific knowledge needed to provide comprehensive information is insufficient; and it would be unethical to collect such information. Additional efficacy or safety studies may be requested, and there may be restrictions on the setting in which the product is used. Conditional approval is reviewed on an annual basis to reassess the risk–benefit balance and the product can be ultimately authorized.

Some EU member states introduced early access programs (EAPs), which will be described in Chapter 9.

For example, France introduced ATU (Temporary Authorization for Use) which follows a specific framework, the protocol for therapeutic use, to ensure patients' follow-up and efficacy and safety data collection.[3]

Another measure used at the country level is off-label use of prescription of drugs. This is not infrequent and takes place in daily clinical practice. In the aim of managing uncertainty and risk, some countries, such as the Netherlands, Germany, and France, have tried to frame off-label drug use under specific conditions, that is, for research use.

Recently, the EMA has proposed "adaptive licensing" as a new approach to licensing medicines.[4] Adaptive licensing was supposed to bring the requirements of regulators and HTA bodies to a closer alignment than before. Indeed, access to new drugs would be based on Bayesian statistics that are considered the reference statistic method to support decision making. However, the practice has shown high reluctance of HTA bodies and payers to take decisions with immature data sets. In order to engage payers, adaptive licensing has become the Medicines Adaptive Pathways to Patients. This illustrates the will of the EMA to modify the process from a purely regulatory one to one that implies that patients are gaining access to new drugs. The EMA is inclusive of HTA in this process, as well as in another new process called PRIME (priority medicines). It also invites HTA representatives when

TABLE 8.7

Regulatory Uncertainty Management in Selected EU Countries beyond EMA Requirements

	France	The United Kingdom	Germany	Italy	Spain	Sweden	The Netherlands
Restricted approval	√	√√	√	√√	√√	√	√
Risk Management Plans	√		√	√	√	√	√
EAPs	√√	√	√	√	√	√	√
Real-world data collection/ monitoring registries	√√		√	√√	√	√√	√√
Off-label use	√	√	√	√	√	√	√

Note: √ shows the extent to which those measures are used: no √- not used; √ - used sometimes; √√ - used often; EAP, Early Access Program.

it prepares scientific advice in order to gain their perspective. Finally, the EMA has launched the HTA–EMA parallel scientific advice process (see Chapter 5).

Table 8.7 shows the management of uncertainty at the regulatory level in selected EU countries.

8.6.2 HTA Bodies and Payers

HTA bodies and payers use various tools for risk management, but they are not a part of structured processes. They try to address uncertainty using technical tools such as health economic models or pragmatic clinical trials to quantify uncertainty. Managing the risk consists of increased methodological requirements for HTA bodies to ensure validity, robustness, and certainty of the evidence produced by manufacturers and/ or cost containment measures to ensure that they do not overpay for a hypothetical benefit that may not materialize.

Some of the technical tools used for extrapolating efficacy in RCTs to real-life effectiveness are: post hoc analysis, meta-analysis, indirect comparison/meta-regression, effectiveness, and cost-effectiveness modeling all involving sensitivity analysis. For instance, in the United Kingdom, the estimation of effectiveness is assessed by modeling incremental cost-effectiveness ratios (ICERs), and the management of uncertainty in the model is made through sensitivity analyses (probabilistic,

deterministic). Each method carries its own uncertainty, but it reduces in all cases initial uncertainty. Using intuitive tools such as pragmatic analysis consists of listing all factors in a drug's RCT that could contribute to reducing effectiveness over efficacy (e.g., using different dosages, short study duration, inappropriate comparator, inclusion criteria).

If uncertainty still remains after previous analyses, some countries resort to specific MA agreements described in Chapter 6, in order to improve the ICER or decrease the budget impact. Uncertainty of a drug's benefit can be addressed through managing budget impact through financial agreements (price discounts, volume agreements, etc.) or managing uncertainty relating to clinical and cost-effectiveness through outcome-based agreements. These are distinguished in two categories: (1) payment for performance (P4P) where payment is decided at an individual level for responder patients and (2) coverage with evidence development (CED).

When uncertainty/risk cannot be dealt with the previously mentioned measures, HTA bodies/payers will simply resort to the management of formulary by restricting the indication to a specific therapeutic indication or subgroups of patients who are the most likely to benefit, thus reducing uncertainty related to the drug's benefit.

Across EU countries, uncertainty is addressed through different tools (Table 8.8).

TABLE 8.8

HTA Bodies/Payers' Uncertainty Management Tools in Selected EU Countries

	France	The United Kingdom	Germany	Italy	Spain	Sweden	The Netherlands
Budget impact analyses	√√		√√	√√	√√		
Cost-effectiveness analysis	√	√√√	√			√√√	√√√
Financial agreements	√√√	√√√	√√√	√√√	√√√	√√	√√
P4P			√√√	√√√	√√√		
CED	√	√	√√			√√√	√√√
Real-world data collection	√√	√	√	√√	√	√√	√
Restricted reimbursement	√√	√√√	√	√√	√	√√	√√

Note: The number of √ describes the extent to which a tool is used. P4P, payment for performance; CED, Coverage with Evidence Development.

- In the Netherlands and Sweden, cost-effectiveness seems to be the driving force and the instrument used is mainly conditional reimbursement (CED).
- Italy uses payment by result and discount or reimbursement for nonresponders, coupled also to data collection as a part of monitoring registries.
- In the United Kingdom, cost-effectiveness is the driving force; however, it uses mostly discounts that do not require additional data collection to improve the drug's cost-effectiveness.
- In Germany, uncertainty is often managed through discounts.
- In France, financial schemes are mostly price/volume agreements. The appreciation of the benefit of postlaunch studies with specific requirements is still in the early stage, although postlaunch studies are a very common requirement.

8.7 TYPE OF STUDIES REQUESTED BY HTA BODIES/PAYERS TO REDUCE THE UNCERTAINTY

HTA bodies and payer authorities may request various kinds of postapproval studies, in order to manage the perceived risks of financing and/or recommending for use of new pharmaceuticals. Figure 8.2 shows that database analysis and observational field studies are the most commonly requested studies by these authorities.

The kind of study that is requested depends on the kind of uncertainty perceived by HTA bodies/payers. For example, uncertainty about the clinical trial design per se that pertains to the trial inclusion criteria, duration, and choice of comparator requires that the manufacturer runs another controlled trial (Figure 8.3). More indirect issues related, for example,

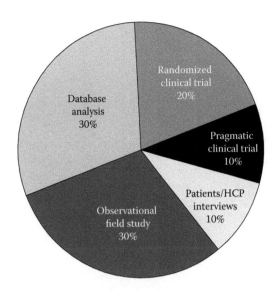

FIGURE 8.2 The share of various kinds of postapproval studies requested by HTA bodies/payers.

to the validity of endpoint used can be resolved by a large observational trial.

Further, payers' uncertainty about the context in which a new drug will be used in real life can be twofold: known from clinical trials before MAu was obtained, or possible to identify only after the drug has reached the market (Figure 8.4). In the former case, HTA bodies/payers may request a pragmatic trial or an observational study, whereas in the latter one, they may wish to closely follow the drug's real-life performance or the availability of alternative treatments or generic options.

Finally, the whole development plan of a new drug may be questioned by HTA bodies/payers, such as lack of appropriate studies or incorrect choice of population in the studies (Figure 8.5).

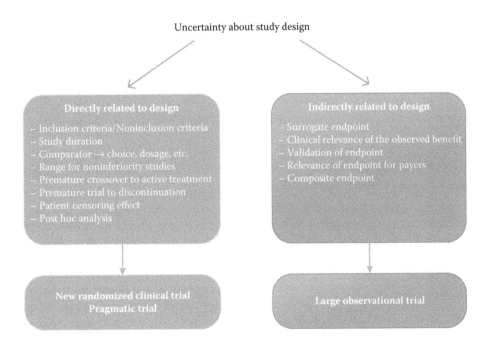

FIGURE 8.3 Trial design-related uncertainties and additional studies requested by HTA bodies/payers.

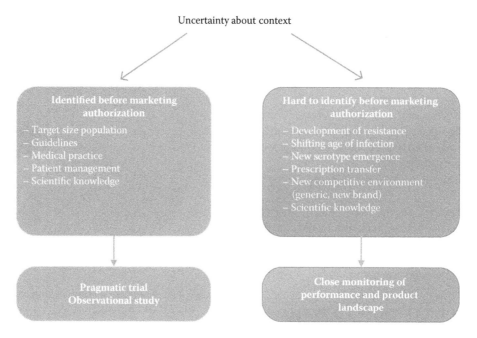

FIGURE 8.4   Context of use-related uncertainties and additional studies requested by HTA bodies/payers.

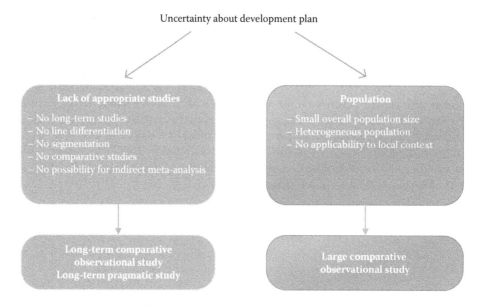

FIGURE 8.5   Development plan-related uncertainties and additional studies requested by HTA bodies/payers.

8.8   CASE STUDIES OF GAPS BETWEEN REGULATORY AND HTA/PAYER APPROVAL

In order to illustrate that gap between regulatory and HTA/payer approvals, we selected five drugs assessed between 2010 and 2012.[5] They are described in Section 8.8.1 through 8.8.3.

8.8.1   DRUGS GENERAL CHARACTERISTICS AND APPROVAL HISTORY

Five drugs that obtained MAu between 2010 and 2012 were analyzed at regulatory (EMA) and HTA/payer levels (Table 8.9).

8.8.2   REGULATORY VERSUS PAYERS' EVALUATIONS

The analysis of the five case studies showed that, generally, the EMA despite the evidence that was not compelling accepted uncertainty and considered the potential benefit of the drug to be outstanding enough to justify MAu, provided that the manufacturers ran specific additional clinical research (Table 8.10), whereas MA decisions at the HTA level were more restrictive, depending on the country (Table 8.11). Generally, payers required harder evidence than regulators, evidence of benefit for primary endpoints such as overall survival, comparative data with adequate comparator, robustness of methodology, effect size, and transferability to clinical practice.

TABLE 8.9
Selected Drugs' General Characteristics

Drug	Pixantrone Dimaleate	Pirfenidone	Abiraterone Acetate	Ipilimumab	Ofatumumab
Marketing authorization holder	CTI Life Sciences Ltd. (Uxbridge, UK)	InterMune UK Ltd. (London, UK)	Janssen-Cilag International N.V. (Beerse, Belgium)	Bristol-Myers Squibb Pharma EEIG (Uxbridge, UK)	Glaxo Group Ltd. (London, UK)
Marketing authorization date	May 10, 2012	February 28, 2011	September 5, 2011	July 13, 2011	April 19, 2010
Indication	Indicated as monotherapy for the treatment of adult patients with multiply relapsed or refractory aggressive non-Hodgkin B-cell lymphomas. The benefit of pixantrone treatment has not been established in patients when used as the fifth-line or greater chemotherapy in patients who are refractory to the last therapy	Indicated in adults for the treatment of mild to moderate idiopathic pulmonary fibrosis	Treatment of metastatic castration-resistant prostate cancer in adult men who are asymptomatic or mildly symptomatic after failure of androgen deprivation therapy in whom chemotherapy is not yet clinically indicated Treatment of metastatic castration-resistant prostate cancer in adult men whose disease has progressed on or after a docetaxel-based chemotherapy regimen	Indicated for the treatment of advanced (unrespectable or metastatic) melanoma in adults who have received prior therapy	Treatment of chronic lymphocytic leukemia in patients who are refractory to fludarabine and alemtuzumab
Conditional approval	Yes	No	No	No	Yes
Approval under exceptional circumstances	No	No	No	No	No
Orphan drug	No	Yes	No	No	Yes
Oncology drug	Yes	No	Yes	Yes	Yes

TABLE 8.10
Regulatory Evaluations by the EMA of the Case Studies

Drug	MAu Date	Therapeutic Area	Orphan Drug	Uncertainty	Risk Management Tool
Ofatumumab	2010	Oncology	Yes	Evidence on positive benefit–risk	Conditional approval/*ad hoc* study
Pirfenidone	2011	Pneumology	Yes	Safety: unfavorable effects	Risk Management Plan: PASS
Pixantrone	2012	Oncology	No	Efficacy: pending results from study PIX 306	Conditional approval
Ipilimumab	2011	Oncology	No	Benefit–risk balance of a higher dosage (10 mg/kg)	*Ad hoc* study
Abiraterone acetate	2011	Oncology	No	Efficacy: overall survival/early study completion	None

Note: PASS, postauthorization safety study.

In countries such as France and Germany even though the drug was accepted for all populations, the benefit of the drug was not always favorably perceived (Table 8.12).

For instance, as an orphan oncology drug targeting a life-threatening disease, ofatumumab was approved by the EMA based on less robust efficacy evidence. However, the drug was granted a conditional authorization because there was a need to further confirm the positive benefit–risk in the double refractory in fludarabine refractory, bulky lymphadenopathy population. The EMA also expressed the need to further confirm the high response rate and control of the disease in the refractory setting through controlled trials and extended ofatumumab treatment. As such, the EMA requested to conduct a phase III study as well as phase IV observational study.

Similarly, pixantrone dimaleate has received a conditional market authorization by the EMA pending the results from Study PIX 306 to support the efficacy of pixantrone dimaleate in patients that had received prior rituximab therapy, as

TABLE 8.11

MA of Drugs at the Country Level

Drug	France	The United Kingdom	Germany	Sweden	The Netherlands	Italy	Spain
Ofatumumab	√	X	NA	NA	√	√ FA	X
Pirfenidone	√	R	√	R, OA	√ FA	√	√
Pixantrone	√	X	√	NA	NA	NA	√
Ipilimumab	√ FA	√	√	NA	√	√	√
Abiraterone acetate	√FA	√	√	NA	NA	√ OA	√

Notes: √, accepted for the whole licensed population; R, accepted for a restricted population; X, not accepted; FA, financial agreement; OA, outcome-based agreement; NA, HTA report not available or not assessed.

TABLE 8.12

HTA Assessments in France and Germany

Drug	France (CT)*	Germany (G-BA)†
Ofatumumab	Moderate SMR‡/ASMR‡ 5	NA+
Pirfenidone	Low SMR‡/ASMR‡ 4	Nonquantifiable additional benefit
Pixantrone	Low SMR‡/ASMR‡ 5	No evidence of additional benefit
Ipilimumab	Important SMR‡/ASMR‡ 4	Considerable additional benefit
Abiraterone acetate	Important SMR‡/ASMR‡ 4	Considerable additional benefit

Notes: CT, Transparency Committee (*Comité de Transparence*); G-BA, Federal Joint Committee (*Gemeinsamer Bundesausschuss*); SMR, Actual medical benefit; ASMR, added actual medical benefit; NA, Not available.

in Europe most patients are expected to have received prior rituximab in this setting.

Evaluation by HTA agencies was in contrast more severe. Both ofatumumab and pixantrone dimaleate were not recommended in the United Kingdom.

In France, ofatumumab was granted a moderate *Service Médical Rendu*/Actual Medical Benefit (SMR) and *Amélioration du Service Médical Rendu*/Added Actual Medical Benefit (ASMR) 5, whereas pixantrone dimaleate was granted a low SMR and ASMR 5.

In Germany, no evidence of additional benefit was concluded by the Federal Joint Committee (*Gemeinsamer Bundesausschuss* [G-BA]) for pixantrone dimaleate.

In Spain, ofatumumab was not reimbursed, and in Italy, it was reimbursed under a cost-sharing MA agreement.

For ofatumumab, HTA agencies criticized the following:

- Efficacy and safety data come from a noncomparative phase II study (France)
- Nonrobust methodology –> difficulty to evaluate effect size (France, the United Kingdom)
- Overall survival was not the primary endpoint (the United Kingdom, France)

For pixantrone dimaleate, HTA agencies criticized the following:

- Lack of information regarding the efficacy of the drug in patients that had received prior rituximab therapy in addition to other elements that weighed in their assessment/HTA conclusions
- Lack of significance in overall survival (OS); only progression free survival (PFS) was significant (France, the United Kingdom, Germany)
- Doubt about generalizability of clinical trial data into clinical practice (France, the United Kingdom)
- Target population was considered small (France, Germany)
- In Germany, there was no evidence of additional benefit because the comparator therapy was not approved in the indication

The EMA approved pirfenidone in its indication under the provision of putting in place a PASS as the Committee for Medicinal Products for Human Use (CHMP) expressed uncertainty around the unfavorable effects.

In contrast, HTA agencies had reservations around additional elements with different evaluations/actions put in place:

- All HTA agencies have overall criticized the lack of use of OS as the main endpoint to evaluate the drug's efficacy.
- The United Kingdom, France, and Sweden have restricted the use of the drug according to specific patient criteria (e.g., forced vital capacity [FVC] below 80%).
- The United Kingdom mentioned the issue of transposability to its patient population and had reservations about the economic model.
- Germany concluded on undefined additional benefit due to nonsignificant results versus best supportive care (BSC).
- Sweden restricted the use to patients with FVC below 80% and requested a MA agreement: comparison of ASCEND study results and updated results from the RECAP study to the values used in the health

economic model, and conditions of use in real clinical practice and how the restriction is followed.

- In the Netherlands, the *College voor Zorgverzekeringen/ The National Health Care Institute* (CVZ) recommended the drug under the condition of a financial agreement.

In the case of ipilimumab, even though the evaluation of the drug was favorable at the EMA level and the HTA level, it is interesting to note that there were differences with regard to the criteria weighing in the decision and the source of uncertainty between EMA and HTA bodies, but also across HTA bodies.

For this drug, the source of uncertainty for the CHMP was regarding the benefit–risk balance of using a higher dosage (10 mg/kg).

In France (ASMR 4), the Transparency Committee (*Comité de Transparence* [CT]) criticized the comparator (gp 100 molecule), which does not have an MAu in the specified indication, the lack of comparative data with vemurafenib, and the robustness of methodology.

In the United Kingdom, the NICE was satisfied that ipilimumab met the criteria for being a life-extending end-of-life treatment and that the trial evidence presented for this consideration was robust. The drug was considered to be cost-effective when applying a Patient Access Scheme (PAS).

In Germany, the G-BA conclusion was a considerable added benefit versus adequate comparator (BSC). The use of gp 100 as a comparator was accepted due to its lack of effect. The G-BA commented, however, on the lack of comparative data with vemurafenib, which has an MAu in this indication.

In the Netherlands, there was uncertainty around the 1-year survival rates of melanoma patients which remains difficult to assess, which efficiency indicator is to be considered to evaluate the magnitude of effect of ipilimumab, and as stated by the EMA, which dose will eventually be used: the registered 3 or 10 mg/kg resulting in cost 3 times as high.

In the case of the drug abiraterone acetate, it is interesting to note the different source of uncertainty across HTA agencies.

In France, the drug received an SMR "important" and ASMR 4. The CT considered that the expected impact on reducing mortality was difficult to assess because the median overall survival was not reached at the time of the second interim analysis and the absence of statistical significance (hazard ratio = 0.752, 95% confidence interval [0.606, 0.934], $p = .0097$ to the limited set of .00008), and when the third interim analysis was performed after switching treatment allowed patients with documented progression of the disease.

However, it is important to note that the clinical trial was stopped by the steering committee because of the important decrease in the mortality rate. All the patients were then treated by abiraterone acetate. Because the study was stopped earlier than expected, the results on progression-free survival were not statistically significant. Even though secondary endpoints were positive, and despite the lack of approved active treatment for this indication, the drug was given ASMR 4.

In Germany, it was the uncertainty regarding harm, and not the benefit regarding the efficacy, which made the The Institute for Quality and Efficiency in Healthcare (*Institut für Qualität und Wirtschaftlichkeit im Gesundheitswese* [IQWiG]) derive a hint of a considerable added benefit of abiraterone in comparison with watchful waiting.

In the United Kingdom, Sweden, and the Netherlands, the indication in patients who have not received chemotherapy has not been assessed.

In Spain and Italy, abiraterone acetate obtained reimbursement, whereas in Italy, a P4P scheme was put in place.

8.8.3 DISCUSSION OF CASE STUDIES

Systematically, the case studies illustrate a gap in uncertainty/risk management between EMA and HTA bodies. There is an opposition between the EMA that constantly weighs the benefit and the risk of granting or not an MAu, and the payer who is reluctant to accept uncertainty/risk and is not willing to pay, and often also not willing to implement risk management tools. Therefore, delayed market access is caused by uncertainty and risk.

There is also a large variability in decisions among HTA bodies. Overall, French HTA body is frequently risk-averse unlike the British and German HTA bodies who are more open depending on the cases. Italy tends to implement systematically P4P (e.g., abiraterone acetate) or financial agreements (e.g., ofatumumab). Ipilimumab is a good example illustrating the contrast between the EMA granting an MAu while requesting further research and the various evaluations across HTA bodies delaying access.

The divergent decisions between regulators and HTA bodies/payers can largely be attributed to the gap in their attitude toward uncertainty and risk in their management strategies. For the EMA, public health is the central concern. When access to an innovative therapy is delayed, the patients' loss of opportunity is weighted against potential safety issues. Therefore, regulators constantly assess the risk–benefit balance of a drug and accept uncertainty as a part of this evaluation. With the unique know-how they have developed, they evaluate the extent of uncertainty and risk using tools that allow them to identify and interpret even early evidence. Uncertainty acceptance is balanced by the development of well-established and standardized tools for management of medical risk in pre- and postapproval phases.

However, HTA bodies and payers are not accustomed to uncertainty and are increasingly risk averse. In spite of using risk-management tools, such as risk-sharing agreements and CED with or without escrow agreements, HTA bodies/payers mainly seek to optimize the use of their budgetary envelope. In practice, the risk management tools are typically sought in view of cost containment, diverting them from their original purpose.

De facto, payers' risk management experience is limited to few emblematic examples, whereas cost containment strategies have developed dramatically. Therefore, unlike regulators, payers discard the patient's loss of opportunity from the decision-making process and only consider high level effectiveness evidence which ensures that they only pay for actual medical benefit. Most payers in Europe adopt this approach because it efficiently reduces the budget impact of a new technology. Within this context, the patient's loss of opportunity is not considered in the decision process. Indeed, the payers are willing to pay for a guaranteed benefit and will only consider hard proof evidences in the decision analysis framework which they had predefined for the HTA process.

Although the divergent decisions between HTA bodies/payers and regulators seem to lead to an impasse solutions to break the deadlock actually exist. On the one hand, the three key actors in the European drug market, that is, the EMA, HTA bodies/payers, and manufacturers, should work toward better collaboration. In particular, the EMA and HTA bodies/payers have to improve their cooperation, which is already in place, in order to better align their decisions for the benefit of patients and society. HTA bodies from various member states and manufacturers should establish early dialogs to better understand HTA expectations (e.g., EUnetHTA), particularly concerning transposability of clinical studies' results in real life. On the other hand, a higher use of tools such as risk-sharing schemes, P4P, CED, and escrow agreements intended to manage risk rather than to contain costs, would have a positive impact on patients' access to innovative therapies. Although payers seem to have fallen behind in the management of uncertainty and risk, positive trends are emerging, especially in Sweden and the Netherlands where payers have been increasingly seeking to implement CED in order to manage the risk.

8.9 CONCLUSIONS

Lack of transferability of clinical trial data to real life has become the primary reason for negative HTA decisions. This fact should be anticipated when phases II and III of clinical trials are designed by manufacturers. Early dialog with HTA bodies is helpful to inform such design.

The development of risk management strategies allows the product to rapidly obtain MA. However, payers' aversion to risk is still a hurdle for MA of new drugs, even with the use of some instruments such as conditional reimbursement (CED). It is likely that increasing cost containment measures will increase payers' aversion to risk and reduce willingness to reimburse products characterised by effectiveness uncertainty.

Without a real engagement in efficient collaboration and the development of risk management strategies, the gap between payers and regulators will continue to increase, aggravating the delay of patients' access to innovative therapies.

REFERENCES

1. Frank, H. K. *Risk, Uncertainty, and Profit*. Boston, MA: Hart, Schaffner and Marx; Houghton Mifflin, 1921.
2. EMA. Risk Management Plans. Available on http://www.ema.europa.eu/ema/index.jsp?curl=pages/regulation/document_listing/document_listing_000360.jsp (Last accessed on December 6, 2013).
3. ANSM. The compassionate use of medicinal products. An example: The French ATU system, March 2012. Available on http://ansm.sante.fr/var/ansm_site/storage/original/application/ebc8e366f2fedd9cc6c32ccdc4e0faa9.pdf (accessed September 29, 2016).
4. EMA. Adaptive Licensing: A useful approach for drug licensing in the EU? Presentation London, March 2012. Available on http://www.ema.europa.eu/docs/en_GB/document_library/Presentation/2012/04/WC500124930.pdf (accessed September 29, 2016).
5. Toumi, M., J. Zard, A. Kornfeld, and C. Rémuzat. Gap between payers and regulators management of risk prevents and delays patient access to new therapy. *ISPOR 19th Annual International Meeting*, Montreal, Canada, May 31–June 4, 2014. PHP31.

9 Early Access Programs

9.1 OVERVIEW

Early access programs (EAPs) are country-specific regulatory processes that grant market access to unlicensed medical drugs to specific patients, under specific terms, provided that they fulfill certain criteria.

Different terms are used for these programs globally. The European Union (EU), through the European Regulation 726/2004/EC, defines compassionate use as a treatment option that allows the use of an unauthorized medicine. Compassionate use programs (CUPs) are for patients in the EU who either have a disease for which no satisfactory authorized therapies exist or cannot enter a clinical trial. They are intended to facilitate the availability to patients of new treatment options under development.[1] The EU guidance is nonbinding, and CUPs are governed by individual EU member states. This led to differences in the interpretation and implementation of the EU regulation, including terminologies. Outside Europe, names and definitions are more variable.

In the United States, Food and Drug Administration (FDA) regulations have allowed patients access to investigational drugs and biologics through expanded access since 1987. Expanded access, also called "compassionate use," is a regulation that makes promising drugs and devices available to patients with serious or immediately life-threatening diseases. The FDA currently approves expanded access on a case-by-case basis for an individual patient, for intermediate-size groups of patients with similar treatment needs who otherwise do not qualify to participate in a clinical trial, or for large groups of patients who do not have other treatment options available and sufficient information is known about the safety and potential effectiveness of a drug from ongoing or completed clinical trials. Just as in clinical trials, these investigational drugs/devices have not yet been approved by the FDA, and they have not been proven to be safe and effective.[2]

Amidst the differences in terminologies, we prefer to refer to these programs as early access programs, and this term will be used for the rest of this chapter. The term, "early access program" may not be found in specific legislations, but it has a wider scope than "compassionate use" and is inclusive of the different programs present globally.

Although there are differences in EAPs among countries, all comply with the following general principles[3]:

1. The primary objective is to provide severely ill patients with early access (i.e., before marketing authorization [MAu] is obtained) to medicinal products.
2. EAPs are distinct from a clinical trial and cannot be used for investigational purposes or commercial preauthorization activities.
3. Normally, EAPs are conducted in parallel with clinical trials of the medicinal product or when the medicinal product is already under assessment for MAu by a competent authority or is already approved but, not yet launched on the market. As such, EAPs may be initiated as early as phase II/III trials, but can start at any time during the approval process or after being granted MAu but prior to market launch, a period that may take 1–2 years. As sufficient safety and efficiency data are needed for the approval of these programs, application is more feasible late in the drug development. However, under exceptional circumstances, drugs may obtain early access with very little clinical information such as for Ebola or H1N1 pandemic risk.
4. EAPs are distinct from off-label use or extension of indication schemes (e.g., Temporary Recommendation for Use [RTU] in France and Medical Needs Programme in Belgium) where a licensed medicinal product is prescribed for an indication different to that for which it was authorized. In EAP, the medicinal product is not licensed in the country and has not been used as a treatment for any disease.[4]
5. EAPs are different to adaptive licensing/fast-track procedures, sometimes known as "staggered approval" or "progressive licensing" (e.g., European Medicines Agency conditional MAu and Notice of Compliance with conditions [NOC/c] in Canada), which is the early authorization of a medicine in a restricted patient population, followed by iterative phases of evidence gathering and the adaptation of the MAu to allow broader patient populations to access the medicine. Adaptive licensing builds on regulatory processes already in place such as EAPs.[5]
6. EAPs are an exceptional and temporary measure, and approval is granted when the following conditions are met:
   a. The medicinal product is intended for the treatment, prevention, and diagnosis of a serious or rare condition.
   b. There is an absence of any suitable therapeutic alternative medicinal product in the concerned country.
   c. The benefit/risk ratio of the medicinal product is strongly presumed to be positive.

Early access to highly needed and innovative drugs is clearly valuable and benefits various health-care stakeholders.

For physicians and patients, EAPs provide access to a treatment option for a life-threatening disease or a serious condition when no approved treatments are available.

For regulatory bodies and payers, EAPs offer the opportunity to evaluate real-world effectiveness and safety to complement randomized controlled trial data and gain more insight into the value of a drug. Moreover, the data collected during the EAP will serve to reassure the pricing committee on the definition of the target population, the benefit(s), usage, and misuse of the drug.

For pharmaceutical companies, the data generated from EAPs provides a base of evidence for safety (at a minimum) and efficacy (where permitted). The experience gained from EAPs would lead to launch preparedness and optimal operational processes. Finally, EAP price may be used for reference in negotiations about subsequent pricing and reimbursement.

9.2 TYPES OF EAPs: NOMINATIVE AND COHORT

EAPs can be divided into two main types of programs: nominative and cohort.

Nominative or named patient EAPs are typically initiated by physicians for an individual patient in great need of a medicinal product, which will be administered under the physician's responsibility. Companies usually have little influence on this type of EAP. However, companies can try to anticipate these demands and define in advance a set of criteria allowing safe access and administration to patients.[3]

Cohort EAPs are usually defined programs initiated by the manufacturer to allow access for a group of patients to an unauthorized medicinal product.[3]

Different countries may refer to them differently, but all the programs fall within this binary classification. The regulatory requirements for each program also vary.

9.3 GLOBAL EAP TRENDS

Tables 9.1 and 9.2 provide a summary of the EAPs in chosen countries:

- Majority of the countries have both nominative and cohort EAPs (France, Italy, Spain, Denmark, Norway, Brazil, and South Korea). The United Kingdom, Switzerland, Australia, Israel, and Turkey have nominative programs only, and only Germany has a cohort program.
- All programs are under the remit of relevant government health authorities.
- There is no pattern in terms of authorization duration and renewability. Some countries have 1-year authorization duration, some have variable duration, and the rest varies from 60 days to 5 years, until the unmet need persists, or until MAu has been granted. The possibility of renewability also varies.
- Commercial provision of drugs in EAP is possible in most of the programs, and the price is usually set freely. In the remaining cases, the price is negotiated with relevant authorities. Reimbursement is usually

conditional. Full reimbursement is only possible in France, Italy, Spain, and in the license procedure in Sweden.

9.4 KEY SUCCESS FACTORS AND EAP MANAGEMENT

Based on the principles found in the EU regulations and national laws, early access can be granted when:

- The patient requiring treatment has a serious or life-threatening disease.
- The patient cannot be treated satisfactorily by an authorized drug or is suffering from a disease where no medicine has been authorized.
- The patient cannot take part in an ongoing clinical trial to obtain a treatment.
- There are sufficient clinical data to permit the evaluation of the benefit versus risk profile of the drug, and there is adequate evidence that the patient will benefit from the treatment.

The more serious the condition and the less effective the existing treatments, the more likely there will be opportunities for EAP. As such, the presence of an unmet medical need is a key factor.

When pharmaceutical companies are considering implementing an EAP, the following should be taken into consideration for success:

- The national regulatory framework and requirements should be considered carefully as there is diversity in the regulatory requirements. This can result in explicit or implicit country-specific barriers to the successful implementation of EAPs. Critical elements of pricing strategy, including price determination, reimbursement possibilities, and the likely payers, must be considered when deciding whether to implement an EAP.[3]
- EAPs need to be well controlled. It is valuable to create an EAP protocol with clear guidelines on the patient inclusion and exclusion criteria, drug use, tracking and safety reporting, and supply chain logistics such as the manufacturing and supply of adequate drugs. The EAP protocol allows for a streamlined implementation of EAPs, the maximization of data collection, and better planning of resources.
- The timing of EAPs is also critical to the success of the program. Application for EAPs is usually feasible late in the drug development, because sufficient safety and efficacy data are needed for the approval of these programs.
- Planning the time of EAP launch should be done early into the drug development process and involve all stakeholders. The timing of EAP initiation should factor regulatory requirements such as the preparation of the EAP dossier and other relevant

TABLE 9.1

EAPs in European Countries

Country	Name of Program	Cohort or Nominative	Government Agency Responsible	Application Initiated By	Authorization Duration and Renewal	Sales	Drug Pricing	Reimbursement	Payer
France	Nominative *Autorisation Temporaire d'Utilisation* (ATU)	Nominative	*Agence Nationale de Sécurité du Médicament et des Produits de Santé* (ANSM)	Physician	One year maximum, renewable	Yes	Free pricing; difference of ATU vs. negotiated price must be reimbursed	100%	National health insurance fund
	Cohort ATU	Cohort	ANSM	Holder of the licensing rights	One year, renewable	Yes	Free pricing; difference of ATU vs. negotiated price must be reimbursed	100%	National health insurance fund
Germany	CUP	Cohort	Federal Institute for Drugs and Medical Devices (BfArM) and the Paul-Ehrlich-Institut (PEI)	Responsible person (pharmaceutical company)	One year, renewable	No	Not applicable	Not applicable	Pharmaceutical company
Italy	Special use: Law 648/96	Cohort	*Agenzia Italiana del Farmaco* (AIFA)	Key Opinion Leaders (KOLs), scientific societies, patient associations	Until the need is persistent and further orders from AIFA	Yes	Free pricing	100%	National Health Service (NHS) through "Fondo AIFA 5%"
	DM 08/05/2003	Nominative and cohort	AIFA	Prescribing physician	Information not available	No	Not applicable	Not applicable	Pharmaceutical company
Spain	Individual access authorization	Nominative	*La Agencia Española de Medicamentos y Productos Sanitarios* (AEMPS)	Hospital	Transitional provisions only but duration and renewability not specified	Yes	Price negotiated with the Interministerial Commission for Pricing of Medicinal Products (CIPM)	100%	NHS
	Temporary use	Cohort	AEMPS	Pharmaceutical company	Transitional provisions only but duration and renewability not specified	Yes	Price negotiated with CIPM	100%	NHS
The United Kingdom	Early Access to Medicines Scheme (EAMS)	Nominative	Medicines and Healthcare products Regulatory Agency (MHRA)	Pharmaceutical company	One year, renewable	No	Not applicable	Not applicable	Pharmaceutical company
	Supply of unlicensed medicinal products ("specials")	Nominative	MHRA	Doctors, manufacturers, etc.	Information not available	Yes	Free pricing	Yes, based on Part VIIIB of the Drug Tariff	NHS

(Continued)

TABLE 9.1 (*Continued*)
EAPs in European Countries

Country	Name of Program	Cohort or Nominative	Government Agency Responsible	Application Initiated By	Authorization Duration and Renewal	Sales	Drug Pricing	Reimbursement	Payer
Denmark	Compassionate use permit	Nominative and cohort	Danish Health and Medicines Authority	Physician	Five years maximum, renewable	Yes	Free pricing	Conditional	Regions via state subsidy
Norway	CUP	Cohort	Norwegian Medicines Agency	Pharmaceutical company	Variable, renewable	Yes	Free pricing	Conditional	National insurance
	Named patient/ approval exemption	Nominative	Norwegian Medicines Agency	Physician or hospital institution	One year, information on renewability not available	Yes	Free pricing	Conditional	National insurance
Sweden	CUP	Cohort	Medical Products Agency (MPA)	Pharmaceutical company	Until MAu or if the company cancels it	No	Not applicable	Not applicable	Pharmaceutical company
	License procedure	Nominative	MPA	Physician, pharmacy	One year, if not stated shorter by MPA	Yes	Pricing can be regulated by TLV	100% but currently under review	County
Switzerland	Special permit (Sonderbewilligung)	Nominative	Swiss Agency for Therapeutic Products (Swissmedic)	Doctors/dentists/ pharmacists	Information not available	Yes	Free pricing	Conditional	Health insurance

TABLE 9.2
EAPs in Non-European Countries

Country	Name of Program	Cohort or Nominative	Government Agency Responsible	Application Initiated By	Authorization Duration and Renewal	Sales	Drug Pricing	Reimbursement	Payer
Australia	Special Access Scheme	Nominative	Therapeutic Goods Administration (TGA)	Physician	Variable	Yes	Free pricing	No	Patient
	Authorised Prescribers	Nominative	TGA	Physician	Variable	Yes	Free pricing	No	Patient
Brazil	CUP	Nominative	Agência Nacional de Vigilância Sanitária; ANVISA	Physician	Information not available	No	Not applicable	Not applicable	Pharmaceutical company
	Expanded Access Program	Cohort	ANVISA	Pharmaceutical company	Information not available	No	Not applicable	Not applicable	Pharmaceutical company
Canada	Special Access Program (SAP)	Nominative and cohort	Health Canada	Physician	Six months, renewable	Yes	Free pricing	Conditional	Patient or hospital, or insurance plan
Israel	Compassionate Use	Nominative	State of Israel Ministry of Health	Physician or medical institution	Variable	Yes	Price negotiated with HMO	Conditional	Health Maintenance Organization (HMO) (Kupot holim)
South Korea	Treatment Use of an Investigational New Drug	Cohort	Ministry of Food and Drug Safety	Pharmaceutical company	Variable, renewability not available	Yes	Price negotiated with NHIC	Conditional	Korean National Health Insurance (KNHI)
	Emergency Use of an Investigational New Drug	Nominative	Ministry of Food and Drug Safety	Physician	Information not available	Yes	Price negotiated with NHIC	Conditional	KNHI
Turkey	CUP	Nominative	Turkey's Ministry of Health, General Directorate of Pharmaceuticals and Pharmacy (IEGM)	Physician	Valid until MAu	No	Not applicable	Not applicable	Pharmaceutical company

documents, supply chain logistics, and relationship building with physicians, patient groups, and regulatory agencies.

- The termination of an EAP should also be properly managed and "walkaway" points should be considered. In most countries, this is possible but a unilateral termination of the EAP is not allowed. In Turkey, the pharmaceutical company must provide the medical product to the enrolled patients up to the end of their treatment and bear all costs. In Spain, the supply of the drug must be guaranteed until MAu is available.
- Cross-functional teams are needed to handle EAP planning, initiation, implementation, and termination.
- Managing both finances and human resources are also vital aspects. Revenues from EAPs are not guaranteed and are dependent on the country and the type of program. Moreover, some countries require application fees, although most do not. Aside from these, internal resource costs, such as management and human resources, manufacturing and supply, logistics, and costs of setting up a registry for tracking and safety reporting, should also be factored in the EAP budget.
- Pharmaceutical companies should consider that EAPs might not be a valuable option if
  - Safety data from the trials are not sound and early real-world usage might hinder the possibility of obtaining MAu.
  - The difficulties in logistics and in supplying the product will disrupt organization, as the initial investment could be quite substantial.

- Once the drug receives its MAu, there is a risk of delayed reimbursement negotiations and strong price opposition as payers know that patients already have access to the drug and consequently are not exercising pressure for the drug to enter the market.[6]

There is currently no evidence that these programs expedite the speed at which medicines receive MAu. Similarly, EAPs do not guarantee MAu, and there is currently no aggregate evidence showing that EAPs will guarantee reimbursement/coverage after MAu is obtained.

REFERENCES

1. European Medicines Agency. Compassionate Use. Accessed October 2014, from http://www.ema.europa.eu/ema/index.jsp?curl=pages/regulation/general/general_content_000293.jsp.
2. U.S. Food and Drug Administration. Understanding Expanded Access/Compassionate Use. Accessed January 2015, from http://www.fda.gov/ForPatients/Other/ExpandedAccess/ucm20041768.htm.
3. Sou, H. EU Compassionate Use Programmes (CUPs): Regulatory Framework and Points to Consider Before CUP Implementation. *Pharm Med* 2010;24(4):1–7.
4. Whitfield, K. et al. Compassionate Use of interventions: results of a European Clinical Research Infrastructures Network (ECRIN) survey of ten European countries. *Trials* 2010;11:104.
5. European Medicines Agency. Adaptive Licensing. Accesses October 2014, from http://www.ema.europa.eu/ema/index.jsp?curl=pages/regulation/general/general_content_000601.jsp&mid=WC0b01ac05807d58ce.
6. Creativ-Ceutical Internal Database. Data on File (Last accessed on December 19, 2014).

10 Market Access of Orphan Drugs

10.1 DEFINITIONS OF ORPHAN DRUGS

Orphan medicinal products, or "orphan drugs," constitute a class of drugs that have been developed specifically to treat a rare medical condition generally referred to as "orphan disease." The label "homeless or orphan drugs" was first used in the United States by G.P. Provost in 1968 to qualify all categories of medications in which the pharmaceutical industry seemed to have very little interest.[1] As the name suggests, rare diseases occur in a very small population.

Therefore, the presumed nonprofitability of orphan drugs derives essentially from the insufficient number of foreseen users, with respect to the costs involved in their development. Therefore, making orphan drugs profitable may require obtaining high prices for a low number of users.

At present, there is no universal definition of rare disease and it differs among countries.[2]

10.1.1 US DEFINITION

In the United States, an orphan drug is defined in the *Orphan Drug Act* as follows: "Orphan drugs are used in diseases or circumstances which occur so infrequently in the USA, that there is no reasonable expectation that the cost of developing and making available in the USA a drug for such disease or condition will be recovered from sales in the USA for such drugs."[3]

The limit of prevalence for a rare condition in the United States is less than 200,000 persons in the country.

In 1985 and 1990, the definition of orphan drugs was extended to products other than drugs, such as biologics, medical devices, and medical foods.

10.1.2 EU DEFINITION

The EU orphan drugs regulation was implemented almost 20 years after the US regulation.

As defined by the regulation EC No 141/2000,[4] a product can be designated as orphan drug, if

It must be intended for the treatment, prevention, or diagnosis of a disease that is life threatening or chronically debilitating.

The prevalence of the condition in the European Union must not be more than 5 in 10,000, or it must be unlikely that marketing of the medicine would generate sufficient returns to justify the investment needed for its development.

No satisfactory method of diagnosis, prevention, or treatment of the condition concerned can be authorized, or, if such a method exists, the medicine must be of significant benefit to those affected by the condition.

10.1.3 JAPAN

In Japan, orphan drugs designation is granted for a product if it fulfills the following criteria[5]:

The disease for which use of the drug is claimed must be incurable. There must be no possible alternative treatment; or the efficacy and expected safety of the drug must be excellent in comparison with other available drugs.

The number of patients affected by this disease in Japan must be less than 50,000 on the Japanese territory, which corresponds to a maximal incidence of 4 per 10,000.

10.1.4 SOUTH KOREA

In Korea, rare diseases are defined as diseases for which there is no appropriate treatment or alternative medicine, which affect fewer than 20,000 people or disease. The Orphan Drugs Guideline was established in 2003. An exclusive marketing rights for 6 years was stipulated to encourage the research and development (R&D) of orphan drugs. Support measures include medical expense reimbursement and nationally funded research programs with support from the Ministry of Health and Welfare and the Korean Centers for Disease Control and Prevention. The Korean Rare Disease Information Database (http://helpline. cdc.go.kr) and the Korean Organization for Rare Diseases (http://www.kord.or.kr) provide vast information on rare diseases for patients, researchers, pharmaceutical companies, and administrators.[6]

10.2 THE LEGAL FRAMEWORKS FOR LICENSING AND ASSESSMENT OF ORPHAN DRUGS AND DEVELOPMENT INCENTIVES

10.2.1 THE EUROPEAN UNION

The European regulation for orphan drugs (EC No 141/2000) was adopted by the European Parliament and European Council on December 16, 1999.[4]

The use of the term "ultra-orphan" is limited to the United Kingdom. Ultra-orphan drugs are defined as drugs used to treat conditions with a prevalence equal to or less than one case per 50,000 population in the UK.[7]

The regulation was designed to incentivize the pharmaceutical industry to the development and marketing of products for rare disease that they will not develop in normal conditions.

This regulation does not mention a special preferential pricing and reimbursement (P&R) status for orphan drugs. P&R are member states' responsibilities. Differences in access in the different member states have resulted.

Several initiatives were taken by the European Commission in order to

Improve recognition of rare diseases
Strengthen the cooperation
Encourage more research
Support the national plans for rare diseases

EU member states offer some incentives for the development of orphan drugs:

- Protocol assistance
- Developmental and regulatory assistance from the EMA's scientific advice may be requested prior to the submission of the marketing authorization application.
- Marketing approval assistance
- Fast-track assessment
  - This accelerated assessment can be used in Europe not only for orphan drugs but also for products of "major public health interest" to treat "serious, chronically debilitating or life-threatening conditions when other methods are absent or by and large insufficient."
- Lower regulatory fees
  - EMA offers fees reductions for designated orphan drugs; those fees include the following:
    – Protocol assistance initial request
    – Follow-up protocol assistance
    – Preauthorization inspections
    – Initial marketing authorization application
    – Postauthorization activities in the first year after marketing authorization, including annual fee and variations
  - This incentive has severely eroded over time for applicants other than small and medium-sized enterprises.
- Market exclusivity
  - The designated orphan drug benefits from a 10-year market exclusivity. This means that for a period of 10 years, neither the EMA nor the national authority will accept an application of marketing authorization for a similar product for the same therapeutic indication.
  - Each orphan drug designation carries market exclusivity for a specific indication. One drug can have several market exclusivities for several indications.
  - The market exclusivity begins after the marketing authorization. This is why in some cases market exclusivity is superior to a patent as the patent can be awarded at early phases of development and reaches its end before the drug reaches the market.

The granting of marketing authorization for a drug does not mean that the drug is available in all countries of the European Union. The drug will go through the necessary procedures in each country in order to establish reimbursement conditions, and usually its price.

Despite joint efforts, the heterogeneity of approaches between countries makes patients' access to orphan drugs more complex.

Compassionate use is an option for using unlicensed medicine in development in patients for whom no authorized satisfactory therapies exist and who cannot enter clinical trials. In some countries, this can be a valid option for companies who wish to sell an orphan drug that is still in clinical development.

10.2.2 France

In 2005, France was the first EU country that established a national plan for orphan drugs that also included funding provisions. It hosts several European organizations that work in the field of orphan diseases, such as EURORDIS (www.eurordis.org), Orphanet portal (www.orpha.net), and the *Orphanet Journal of Rare Diseases* (www.ojrd.com).

10.2.2.1 Compassionate Use

Known as temporary authorization for use (*autorisation temporaire d'utilisation* [ATU]), this authorization allows products to reach the market before marketing authorization. It is issued exceptionally and temporarily by the regulator *Agence Nationale de Sécurité du Médicament et des Produits de Santé* (ANSM).[8]

There are two types of ATU: cohort and nominative (Table 10.1).

To receive an ATU, the product must meet the following criteria:

- It must be a medicinal product (not an extemporaneous preparation).
- It has no marketing authorization in France (for any indication). If it has a marketing authorization abroad, then this status should be sought for France.
- It offers treatment, prevention, or diagnosis for a serious or rare disease.
- No satisfactory alternative treatment with a marketing authorization is available in France.
- Positive risk/benefit ratio is expected for the patient.
- Treatment cannot be postponed.
- Use and supply is limited to hospitals only. Prescriptions may be restricted to certain specialists.
- No product advertising is allowed.
- The aim is therapeutic (clinical trial approval should be sought for research).

The scheme is particularly popular for orphan drugs, especially when no or few poorly effective alternatives exist.

10.2.2.2 Development Incentives

Companies developing new drugs for orphan diseases can also obtain free scientific advice from the ANSM. Depending on the level of sales at public prices, orphan drugs can receive up to 100% tax breaks.[11]

TABLE 10.1

Comparison of Cohort and Nominative Compassionate Use Programs in France

Cohort ATU (*ATU de cohorte*)	Nominative ATU (*ATU nominative*)
Group of patients, one indication	One patient, on a named-patient basis
Application made by the manufacturer, commitment to submit a future marketing authorisation application	Patient cannot enter a clinical trial, request made by and responsibility lies with physician
Safety and efficacy of the product are highly presumed	Safety and efficacy of the product are presumed
ATU for 1-year duration, renewal possible	ATU for the duration of treatment
Follow-up of all patients and data collection (safety and efficacy) according to a protocol for therapeutic use	Follow-up of all patients and data collection (safety and efficacy) according to a protocol for therapeutic use
Periodic data reporting (including any new data that could impact patients' safety) to ANSM	

Note: ANSM, Agence Nationale de Sécurité du Médicament et des Produits de Santé; ATU, autorisation temporaire d'utilisation.

10.2.3 Germany

In contrast to France, Germany has not developed early national plans to rare diseases. However, this plan is under finalization and should be released soon. An association of stakeholders called the National Action League for People with Rare Diseases was formed in 2010 as a result of slow progress in development of relevant legislation.

10.2.3.1 Compassionate Use

Although compassionate use in Germany must involve free provision of a drug by the manufacturer, off-label use can be reimbursed under certain conditions:

The drug is used to treat a fatal or life-threatening disease.
There is no approved product for the indication available in Germany.
There is scientific evidence of positive therapeutic effects.
Advice on off-label use to the Federal Joint Committee (*Gemeinsamer Bundesausschuss* [G-BA]) is provided on an individual basis by expert groups who decide on listing such drugs.

10.2.3.2 Development Incentives

Special regulation exists in the Health Technology Assessment (HTA) for orphan drugs in Germany. There is an amendment to the Social Code Book (SGB V) that says that no new data need to be submitted on the extent of additional benefit over a comparator for an EU-designated orphan drug with an annual out-of-hospital turnover less than €50 million in the previous 12 calendar months (Section 35a(1) sentence 10 SGB V). For such drugs, the German Institute for Quality and Efficiency in Healthcare (IQWIG) assesses only the number of patients treated and the cost of the orphan drug for the concerned population. If the assessment confirms a turnover lower than the arbitrary value of €50 million, G-BA should acknowledge the added benefit automatically. This regulation is meant to facilitate the reimbursement of orphan drugs and to encourage innovation in the field.

If after launch at any time, the yearly revenues exceed the threshold value, the manufacturer needs to prove an added benefit of the new medicine via the IQWiG assessment pathway.

Drugs that have a turnover higher than €50 million are assessed by IQWiG as other drugs. Lower significance levels for p values (e.g., 10% significance) can be acceptable for early, small-size studies on orphan drugs. However, randomized controlled trials remain the agency's preferred kind of evidence for orphan drugs.

10.2.4 Spain

Spain was the second European country that published a national strategy for rare diseases in 2008. Some regions such as Andalucia, Extremadura, and Catalonia have created their own rare disease plans. Special measures regarding the manufacture, importation, distribution, and dispensing of orphan drugs are provided in Article 2 of Law 29/2006.

10.2.4.1 Compassionate Use

Article 24.4 of Law 29/2006 states: "the prescription and application of unauthorised medicinal products to patients who are not included in a clinical trial, with the objective of satisfying, through a compassionate use, the special treatment needs of the clinical situations of specific patients."[10] Off-label use of authorized medicines and nonauthorized medicines are covered in this article. Applications to import an unapproved product from another country, supported by a medical report, have to be made in advance to the Spanish Medicines Agency.

The drug can be used in the following situations:

A chronic or seriously debilitating or life-threatening disease, which cannot be treated satisfactorily with an authorized product (or one that is authorized but not marketed in Spain).
The product must be subject to an application for marketing authorization or undergoing clinical trial.
The sponsor of the clinical trial or the applicant for marketing authorization must have consented to supply the product.

10.2.4.2 Development Incentives

Typically, a 7.5% rebate is mandatory on the value-added tax (VAT)-exclusive public price of all medicines funded by the National Health Service (*Servizio Sanitario Nazionale* [SNS]) for ambulatory or hospital use, if not included in the reference price system. In the case of orphan drugs, the rebate is limited to 4%.

10.2.5 ITALY

Rare diseases have been considered a public health priority in Italy since 1998.[11] A 2001 Ministerial Decree 279/2001 provided a list of 284 rare diseases and exempted patients with those conditions from copayments in the National Health Service. It also made provision for a national network of centers for the prevention, diagnosis, and treatment of rare diseases.

10.2.5.1 Compassionate Use

Law 94/1998 (Legge Di Bella) permits the use of unapproved drugs on the physician's responsibility. The patient needs to give an informed consent. The Technical Committee of the *Agenzia Italiana del Farmaco* (AIFA) can include a given medication into a special list allowing it to be prescribed at the National Health Service (NHS) charge, if for a specific disease there is no therapeutic alternative. Three types of medical products that can be included are as follows:

- Innovative drugs whose sale is authorized abroad, but not in Italy
- Drugs not yet authorized but which underwent clinical trials
- Drugs to be used for a therapeutic indication other than the one that has been authorized.

Three National Healthcare Plans (1998–2000; 2003–2005; 2006–2008) and Regional Health plans were formulated where rare diseases were addressed.

10.2.5.2 Development Incentives

In 2005, the AIFA established a "Fondo AIFA 5%" scheme, in order to oblige pharmaceutical companies to contribute 5% of their total promotional expenditure (excluding salaries) to a special fund, half of which is used to pay for independent research, drug information programs, or pharmcovigilance on EU-approved or EU-designated orphan drugs by those working in public or nonprofit institutions, and half goes toward reimbursing the cost of unapproved products. This increases the budget available to AIFA to reimburse orphan drugs.[12]

10.2.6 THE UNITED KINGDOM

10.2.6.1 Scotland

The assessment process for orphan drug submissions is the same as for all other drug submissions. However, Scottish Medicines Consortium (SMC) may consider additional factors in addition to the usual assessment of clinical and cost-effectiveness, such as whether the drug treats a life-threatening disease; increases quality of life and/or life expectancy; can reverse, rather than stabilize, the condition; or bridges a gap to a "definitive" therapy.[13]

SMC recognizes that orphan drugs may have a less well-developed clinical trials program and, therefore, that less information than usual may be available for some sections of the application. However, more detail may be required in other areas, for example, validity of surrogate markers.

If there is a significant lack of data on long-term outcome with an orphan drug, this surveillance may include specific clinical audit and, where relevant, a patient register.

SMC does not have a fixed incremental cost-effectiveness ratio (ICER) limit beyond which drugs are not recommended, but ICER serves as one of the factors when making decisions. If ICER is relatively high, the following factors may be considered:

Evidence of a substantial improvement in life expectancy (with sufficient quality of life to make the extra survival desirable). Substantial improvement in life expectancy would normally be a median gain of 3 months, but the SMC assesses the particular clinical context in reaching its decision

Evidence of a substantial improvement in quality of life (with or without survival benefit)

Evidence that a subgroup of patients may derive specific or extra benefit and that the medicine in question can, in practice, be targeted at this subgroup

Absence of other therapeutic options of proven benefit for the disease in question and provided by the NHS

Possible bridging to another definitive therapy (e.g., bone marrow transplantation or curative surgery) in a defined proportion of patients

Emergence of a licensed medicine as an alternative to an unlicensed product that is established in clinical practice in the NHS Scotland as the only therapeutic option for a specific indication

10.2.6.2 England and Wales

A threshold of ICER per quality adjusted life year (QALY) is the benchmark of National Institute for Health and Care Excellence (NICE) recommendations for regular medicine. However, the NICE is developing a new methodology to evaluate ultra-orphan drugs, called highly specialized technology (HST).[14] The NICE defines ultra-orphan drugs as diseases with a prevalence of less than 1000 patients in the United Kingdom or less than 1 in 50,000 people.

The HST criteria are as follows:

The nature of the condition

The impact of the new technology

The cost to the NHS and Personal Social Services value for money

The impact of the technology beyond direct health benefits

The impact of the technology on the delivery of the specialized service

The HST program process starts with a consultation on provisional list of topics. The topics are then referred by the

Minister of Health to the NICE. An evaluation committee considers all evidence from submissions from the manufacturer, patients, clinical specialists, NHS England, evidence review group (ERG) report, premeeting briefing, followed by a public consultation for 4 weeks. At the end of the process, the final evaluation determination is produced and the guidance is issued. This process takes around 27 weeks.

Other considerations of HST are as follows: "When evaluating cost to the NHS and PSS, the Committee will take into account the total budget for specialised services, and how it is allocated, as well as the scale of investment in comparable areas of medicine. The committee will also take into account what could be considered a reasonable cost for the medicine in the context of recouping manufacturing, R&D costs from sales to a limited number of patients."

For ultra-orphan drugs, the NICE would like to operate as a broker putting together all the stakeholders around the same table and looking for a reasonable price that would satisfy all parties. The expected budget impact of such drugs is small or negligible, whereas the value for individual patients may be enormous. Acting as a broker, the NICE would avoid using the ICER value in the decision-making process, which should allow greater patient access for such products.

For drugs used in terminally ill patients, the NICE has a separate guidance called "end-of-life" supplementary advice. As per the rule, higher ICER thresholds are acceptable compared to those for regular drugs. Increasing ICER level enables the NICE to give positive recommendation on such drugs. To qualify, the medicine needs to:

- Be approved for treating a patient population normally not exceeding 7,000 new patients a year
- Have no alternative with comparable benefits available though the NHS
- Be indicated for the treatment of patients with a diagnosis of a terminal illness and who are not, on average, expected to live for more than 24 months
- Have sufficient evidence to indicate that it offers a substantial average extension to life, normally of at least 3 months, compared to current treatments
- Have been assessed by the NICE as having an ICER in excess of GBP 30,000

10.2.7 ASIA

In Asia, orphan drug legislation has been adopted in Japan, South Korea, Singapore, and Taiwan.[15] Incentives are playing an important role in encouraging manufacturers to develop orphan drugs. The current regulations only set forth general criteria to accelerate the registration and approval of orphan drugs, but detailed rules have not been implemented and further incentives have not been proposed until now.[16] In spite of being growing pharmaceutical markets, China and India do not have legislation or a national plan for orphan drugs.

10.2.7.1 Japan

The relevant regulatory authority in Japan is the Ministry of Health, Labor, and Welfare (MHLW). The Orphan Drug Regulation is Pharmaceutical Affairs Law Amendment (1993). There are several incentives in Japan[14,17]:

- The orphan drugs benefit from a 10-year marketing exclusivity[15]: The applicant can be granted a 10-year market exclusivity from generic competition. It is possible that the MHLW may reduce this exclusivity. Product renewal period is 10 years for orphan drugs, compared to 4–6 years for other drugs
- Various tax incentives such as 6% tax credit for any type of study and up to 10% of corporate tax
- Fifty-percent reimbursement of development cost: the National Institute of Biomedical Innovation (NIBIO) offers grants for the development of orphan products up to 50% of the R&D cost
- Regulatory meetings: free advice on the clinical trial design is provided by the NIBIO. Fees associated with application submission are reduced by 25%
- Marketing authorization via fast track is possible in Japan: The orphan drug application is given preference over other products in every step of the review process
- The target review time is 6 months. Average approval time for orphan drugs is approximately 10 months, compared to 12 months for other drugs

10.2.7.2 South Korea

The relevant regulatory authority in South Korea is the Korean Food and Drug Administration (KFDA).[15] The legal framework is contained in the Ministry Of Food And Drug Safety (MFDS) Notification No. 2013/222: Provision on Designation of Orphan Drugs (1998) and Orphan Drug Provision designation of 2013. Orphan drugs receive a fast-track review by the KFDA and a 50% application fee waiver. There is no marketing data exclusivity granted.

10.3 THE PRICING PROCESS OF ORPHAN DRUGS

Below, we present information specific to orphan drugs. It should be complemented by information discussed in sections on pricing in Chapters 12 through 21.

10.3.1 FRANCE

Pricing and reimbursed applications for new drugs are combined and form a three-step process:

The Transparency Commission (*Commission de la Transparence* [CT]), at the National Authority for Health (*Haute Autorité de Santé* [HAS]), evaluates the drug's actual medical benefit (*Service Médical Rendu* [SMR]) and added actual medical benefit (*Amélioration du Service Médical Rendu* [ASMR]) compared to existing therapeutic options. Broadly, SMR influences the reimbursement rate and ASMR the price. Medicines with no benefit are excluded from reimbursement.

The Economic Committee of Health Products (CEPS) negotiates the maximum selling price with the company based on the SMR/ASMR rating, prices in selected European countries, the size of the target population, the sales volume forecast for each of the first 5 years, conditions of use, and prices in France of other drugs in the same therapeutic class.

The National Union of Health Insurance Funds (*Union nationale des caisses d'assurance maladie* [UNCAM]) that provides complementary health insurance to the population can modify the reimbursement level proposed by CT by ±5 points. Finally, a pricing agreement is signed between the CEPS and the manufacturer.

In France, orphan drugs are assessed as regular medicines and according to the aforementioned rules. Most orphan drugs examined by the CT so far have received an SMR "important" (i.e., "irreplaceable and particularly expensive") and thus qualified for complete reimbursement. However, the ASMR rating that affects the price has been variable, and it is not rare that prices are reduced following a scheduled reevaluation of ASMR.

10.3.2 Germany

In Germany, only drugs with proven added benefit qualify for premium price negotiation with health insurers. However, the country has a special assessment pathway for orphan drugs. In contrast to regular medicines whose added benefit needs to be proven by manufacturers and confirmed by IQWiG, it is automatically considered as proven for orphan drugs whose yearly turnover with the statutory health insurance does not exceed €50 million in two consecutive years. Drugs with automatically proven benefit qualify for premium price negotiations. The final price is negotiated between manufacturers and the association of statutory health insurances.

However, even though there is no added benefit assessment by IQWiG for drugs whose turnover is less than the threshold value, G-BA performs its own assessment, based on a "lean dossier" prepared by the company that includes data used to license the drug and relevant supplements.

Certain orphan drugs with limited clinical evidence have obtained acknowledgement of benefit, but sometimes it was considered as "nonquantifiable." These appreciations can obviously affect the price negotiation between the association of sickness funds and the company.

Another category of drugs exempted from benefit assessment by IQWiG are drugs with "insignificant" out-of-hospital expenditure for the statutory sickness funds (sales less than €1 million over 12 months). The company must submit a petition to the G-BA to apply for the exemption.

10.3.3 Italy

The AIFA needs to agree on the maximum prices of all medicines seeking reimbursement under the NHS (SSN) and those drugs are generally reimbursed at 100%. The pricing consist in an informal negotiation between the manufacturer and the AIFA and may involve confidential discounts and rebates.

10.3.4 Spain

Pricing and reimbursement decisions are combined and managed by the Ministry's General Subdirectorate of Quality of Medicines and Health Products (*Subdirección General de Calidad de Medicamentos y Productos Sanitarios*), but its recommendations are often ratified by the Interministerial Commission on Medicines Prices (*Comisión Interministerial de Precios de los Medicamentos*), which also takes the final decision.

Severity of the disease, the unmet medical needs, the therapeutic and social utility of the medicine, the degree of innovation, prices in selected EU countries and in the country of origin, prices of similar drugs marketed in Spain, decisions made by HTA bodies abroad (e.g., NICE, IQWiG, HAS), a 3-year sales forecast, and the manufacturer's profit are all factors considered in the pricing decisions. Once products achieve a price and reimbursement at the central level, they need to be listed at the regional level in order to gain market access and be available for patients.

10.3.5 The United Kingdom

Prices of medicines in the United Kingdom are not regulated by any authority, but highly priced medicines that face negative recommendation by NICE may be subject to confidential discounts or rebates agreed with the Department of Health. These are known as Patient Access Schemes (PAS). The price regulation operates indirectly in the United Kingdom through the Pharmaceutical Price Regulation Scheme (PPRS).

10.3.6 Japan

Most new orphan drugs in Japan are priced by cost calculation, but if there is a similar price-listed drug, its price is used as a benchmark. In the latter case, premiums may be added according to the new drug's perceived level of usefulness, innovation, and/or market size. The limited marketability premium applies specifically to Japan-designated orphan drugs priced by cost comparison, but only one product has achieved the maximum 20% premium. The final calculated price is subject to the foreign price adjustment rule. The entire pricing process is relatively rapid, with a target of 60 days (maximum 90 days) for completion. Subsequent listing in the National Health Insurance Tariff is required before prescribing takes place. In part due to the strength of the yen, Japanese orphan drug prices can be among the highest in the world. Nine brands have achieved domestic sales in excess of ¥10 billion/year. Demand-side controls are limited and access to orphan drugs is considered good.

The Japanese National Health Insurance negotiates the prices with the manufacturer after drug approval, allowing a selling price of 'cost plus 10%' for orphan drugs. Almost half of the orphan drugs on the Japanese market originated from the European Union or the United States. Moreover, 56 of 130 designated rare diseases in Japan are subject to reimbursement of medical expenses, with 30% of expenses paid by insurance companies and the rest paid by national and prefectural governments.[18] In 2010, reimbursements expanded to ¥28 billion and the number of recipients expanded to approximately 700,000.[19]

10.3.7 SOUTH KOREA

Orphan and non-orphan drugs follow the same pricing pathway. Reportedly, the resulting price is often disappointing for multinational companies.[6]

10.4 COMPARISON OF PRICES OF ORPHAN DRUGS

Generally, there are no large variations in ex-factory prices for orphan drugs between countries of different pricing and reimbursement systems. Countries with free market pharmaceutical pricing (e.g., UK may have higher public prices for orphan drugs than countries that regulate prices (e.g., Spain).[20] However, due to national budget constraints there may be confidential discounts and rebates in place which obscure the real cost of these drugs to the payers.

10.5 THE HTA FRAMEWORK FOR ORPHAN DRUGS AND ULTRA-ORPHAN DRUGS

Different jurisdictions focus on various HTA criteria for the evaluation of orphan drugs, such as cost-effectiveness, budget impact, disease severity, therapeutic need, and social benefits. There is no universal HTA framework, and the existing approaches are facing many challenges.

Standard HTA approaches that require data from randomized controlled trials are often relaxed when applied to orphan drugs.[15] This is because there may be little data available, or the data may be of low quality (e.g., small trial patient population, surrogate endpoints used, etc.), even if the drug in question has been licensed for use. Because of the data paucity, higher levels of uncertainty on clinical efficacy, safety, incremental cost-effectiveness, and budgetary impact may be allowed in certain countries. Additionally, other criteria may be taken into account in HTA analyses, such as therapeutic value, budget impact, impact on clinical practice, pricing and reimbursement practices globally, patient organizations, economic importance, ethical arguments, and the political climate.[15] Similarly, although cost-effectiveness modeling is feasible for ultra-orphan drugs, they will typically not meet the criteria for cost-effectiveness.[21]

However, these various approaches result in disparities in access to orphan drugs among countries. For example, the United Kingdom's NICE gave only two positive recommendations on 43 EMA-approved orphan drugs, 69% of them were reimbursed in Sweden, and 94% and 100% of them were reimbursed in Italy and France, respectively.[15]

Interestingly, France and Italy focus on criteria such as proven clinical value, evidence from cohort studies, and the degree of innovation, but do not require a formal cost-effectiveness analysis for orphan drugs. In spite of the high price of orphan medicines, they are reimbursed in these countries because of their relatively low budget impact due to small patient populations. In contrast, the United Kingdom focuses on cost-effectiveness and estimates of ICER per QALY. This illustrates a trend where countries that require standard cost-effectiveness analysis typically have a lower reimbursement coverage of orphan drugs than countries that do not. Consequently, patients with rare diseases in countries that employ solely the cost-effectiveness approach may be deprived of access to orphan drugs.

10.6 THE CONCEPT OF ETHICS AND EQUITY FOR ORPHAN DRUGS

As shown previously, ICER-based decision making that focuses on allocation of limited resources in order to maximize the health value generated may not be compatible with the pursuit of social equity. The social preferences to value the severity of the condition and the urgency of an intervention in the case of orphan drugs are opposed to the individual preferences to value the capacity to benefit from an intervention in the cost-effectiveness traditional approach.

However, incorporating social values into the HTA framework requires more empirical research that measures the social preferences in a given society. Nevertheless, such survey data can be confounded by the respondents' understanding of consequences of their preferences. For example, one study found that despite a society's desire for equal treatment rights for patients with rare diseases, there was little preference if the treatment of patients with rare disease was at the expense of treatment of people with common conditions.[22] This is because people can share two notions of equity: horizontal equity (equal treatment of equals, implying that everyone in the society is equal by birth and spending health-care budget on rare diseases is unfair) and vertical equity (unequal treatment of unequals, implying that people in the society are not equal by birth [e.g., in terms of their genetic makeup] and therefore are entitled to special treatments).[28] From the utilitarian perspective of allocation of limited resources, funding of orphan drugs must support the vertical equity.

10.7 POTENTIAL ALTERNATIVE METHODS FOR HTA AND PRICING OF ORPHAN DRUGS

Because cost-effectiveness thresholds alone are not adequate to assess the value of orphan drugs, other criteria need to be taken into consideration. Multicriteria decision analysis (MCDA) enables decision makers to explicitly trade off various nonmonetary factors against each other, alongside cost-effectiveness. To apply MCDA, the relative weight given to each factor in a society or decision-making setting needs to be assessed first.

A pilot study on the use of MCDA for orphan drugs proposed eight nonmonetary criteria in two categories[23]:

> Impact of the rare disease and associated unmet need:
> > Availability of effective treatment options/best supportive care in the absence of the new medicine
> > Disease survival prognosis with current standard of care
> > Disease morbidity and patient clinical disability with current standard of care
> > Social impact of the disease on patients' and carers' daily lives with current standard of care
> Impact of the new medicine:
> > Treatment innovation, defined as the scientific advance of the new treatment together with contribution to patient outcome
> > Evidence of treatment clinical efficacy and patient clinical outcome
> > Treatment safety
> > Social impact of the treatment on patients' and carers' daily lives

Interestingly, the authors found that large weight was given by respondents to the nature of the disease being treated, rather than to the result of using the medicine to treat it. This means that the studied population would be willing to value treatments for rare diseases, even if the treatment outcomes were uncertain.

However, many orphan drugs would not be recommended for reimbursement even if MCDA was applied, because of their high prices.

Therefore, methods such as "cost plus" or "rate of return" could be considered when pricing orphan drugs. However, it is complex to assess objectively what is the cost of developing a drug and how to account for the cost of unsuccessful candidate molecules that had to be discontinued without financial return to manufacturers.

Both HTA and pricing methods should take into account that certain orphan drugs obtain multiple indications, often not for rare diseases. There may be a perverse strategy from manufacturers to first license such drugs in the orphan designation in order to obtain a higher price and then obtain non-orphan designations for the same molecule. In such cases, HTA and pricing processes need to be revised in this new indication, and the price and/or reimbursement rate needs to be adjusted. However, it is unclear how to calculate such adjustments, and if they should be applied on all indications or only on non-orphan ones.

Regardless of the methods adopted in the future, it is clear that the assessment of orphan drugs cannot follow the same pathway as that of regular medicines. Otherwise, these drugs are unlikely to be recommended for reimbursement and patients will be deprived access to the necessary treatments.

10.8 THE ISSUES WITH PRICES OF ORPHAN DRUGS

R&D costs for orphan drugs are 25% of the costs of standard drugs, and manufacturers can quickly recover their investment by obtaining a conditional, early approval for the orphan molecule.[24] This is possible because, in spite of the small size of the target population, these molecules have high prices and can be as profitable as standard medicines to the manufacturers. Indeed, prices of orphan drugs often exceed €100,000 per patient per year.

For example, agalsidase alfa for Fabry disease, a rare genetic condition, costs on average US $265,987.20 per patient per year.

Because there are few or no alternative treatments available to treat rare diseases, manufacturers of orphan drugs have the market monopoly. Additionally, both the FDA and the EMA grant manufacturers of orphan drugs extended periods of marketing exclusivity. Further, the negotiating power of payers is limited, often as a result of political pressure to make new treatments available.[22]

Interestingly, our preliminary research on oncology orphan drugs approved in the US has shown that there is no correlation between their prices and the sizes of target populations of patients.[29]

10.9 FUTURE PERSPECTIVES

The current orphan drug policies have accelerated the development and licensing of drugs for rare diseases that would not be delivered otherwise. However, due to the high prices of orphan drugs, these policies are unlikely to be sustainable. According to Cote et al., there are several reasons for that, including the following ones[25]:

- Fast-tracking of approval of orphan drugs by the FDA and the EMA, including free or discounted assistance in creating clinical development plans
- Excessive stratification of indications leading to a three-step strategy: (1) Apply for orphan designation; obtain substantial economic benefits during the development, approval, and marketing phases; and demand a high price because of the low prevalence of the initial target population; (2) after approval, convince doctors to use the drug in their practice; and (3) expand sales by obtaining new therapeutic indications, orphan or otherwise, while maintaining the initial price
- Repurposing obsolete molecules in new orphan indications, which can lower the R&D costs significantly
- Pricing based on willingness to pay, rather than on the true cost of development to the manufacturers
- Off-label prescriptions of orphan drugs for nonrare diseases

A novel policy framework has been proposed by the Ontario Public Drug Programs, which involves understanding the drug's value before calculating its cost-effectiveness.[26] The drug value assessment step includes criteria such as the following:

- Disease improvements are strongly associated with drug exposure.
- Disease improvements have been repeatedly associated with drug exposure despite variations in population, disease stage, and therapy.
- Benefit with the drug is specific to the disease or disease mechanism.
- Disease improvements occur after drug exposure.
- Optimal disease improvements are associated with optimal drug doses.
- Drug mechanism addresses underlying disease pathophysiology.
- Observations of drug effects do not conflict with generally known facts of natural history and biology of the disease.
- Experimental data confirm disease improvement with drug exposure.
- Drugs of similar mechanism improve similar diseases.

The outcomes of modeling are consulted with clinical experts and other stakeholders. The steps are repeated when new information becomes available and the policy recommendations are reviewed. Even though the model assumes the availability of adequate information about disease incidence and natural history, it is based on the most contemporary data and expert opinion available relevant to the disease, and utilizes reasonable expectations of effectiveness at a point in the disease process when these are most likely to be realized.[26]

10.10 CONCLUSION

Orphan drug policy incentives have stimulated the pharmaceutical industries to the development of research into diseases with significant unmet medical need. The revenue-generating potential of orphan drugs is similar for non-orphan drugs, even though patient populations for rare diseases are significantly smaller.[27] Moreover, orphan drugs may be more profitable, when considered in the full context of developmental drivers, including government financial incentives, smaller clinical trial sizes, shorter clinical trial times, and higher rates of regulatory success. However, current orphan drug policies are unlikely to be sustainable, because they have led to high prices of orphan drugs and to limited coverage and restricted patient access when cost-effectiveness is the sole decision-making criterion.

REFERENCES

1. Provost G. 1968. Homeless or orphan drugs. *American Journal of Hospital Pharmacy*, 25:609.
2. Richter T, Nestler-Parr S, Babela R, Khan ZM, Tesoro T, Molsen E, Hughes DA. Rare Disease Terminology and Definitions-A Systematic Global Review: Report of the ISPOR Rare Disease Special Interest Group.
3. FDA, Orphan Drug Act. Available at: http://www.fda.gov/ForIndustry/DevelopingProductsforRareDiseasesConditions/HowtoapplyforOrphanProductDesignation/ucm364750.htm.
4. Regulation (EC) No 141/2000 of the European Parliament and of the Council of 16 December 1999 on orphan medicinal products, OJ L 18, 22.1.2000, p. 1–5.
5. Orphanet, Orphan drugs in Japan. Available at: http://www.orpha.net/consor/cgi-bin/Education_AboutOrphanDrugs.php?lng=EN&stapage=ST_EDUCATION_EDUCATION_ABOUTORPHANDRUGS_JAP#policy.
6. Just Pharma Reports, 2010. Orphan Drugs in Asia-Pacific: From Designation To Pricing, Funding & Market Access.
7. Process for appraising orphan and ultra-orphan medicines and medicines developed specifically for rare diseases, 2015. Available at: http://www.awmsg.org/docs/awmsg/appraisaldocs/inforandforms/AWMSG%20Orphan%20and%20Ultra%20Orphan%20process.pdf.
8. ANSM, Autorisations temporaires d'utilisation. Available at: http://ansm.sante.fr/Activites/Autorisations-temporaires-d-utilisation-ATU/Qu-est-ce-qu-une-autorisation-temporaire-d-utilisation/(offset)/0.
9. Benefit Assessment of pharmaceuticals according to s. 35a SGB V, Available at: http://www.english.g-ba.de/downloads/17-98-3042/Chapter5-Rules-of-Procedure-G-BA.pdf.
10. Law 29/2006 on the guarantees and rational use of medicines and healthcare products, Available at: http://www.boe.es/boe/dias/2006/07/27/pdfs/A28122-28165.pdf.
11. Just Pharma Reports, Orphan Drugs In Europe. Pricing, Reimbursement, Funding & Market Access Issues, 2013 Edition.
12. AIFA, Fondo AIFA 5%. Available at: http://www.agenzia-farmaco.gov.it/it/content/fondo-aifa-5.
13. SMC Modifiers used in Appraising New Medicines. Available at: http://www.scottishmedicines.org.uk/About_SMC/Policy_statements/SMC_Modifiers_used_in_Appraising_New_Medicines.
14. NICE, Highly Specialised Technologies Guidance, Available at: https://www.nice.org.uk/about/what-we-do/our-programmes/nice-guidance/nice-highly-specialised-technologies-guidance.
15. Gammie T, Lu CY, Babar ZU-D. 2015. Access to Orphan Drugs: A Comprehensive Review of Legislations, Regulations and Policies in 35 Countries. *PLoS ONE*, 10(10):e0140002.
16. Song P, Gao J, Inagaki Y, Kokudo N, Tang W. 2012. Rare diseases, orphan drugs, and their regulation in Asia: Current status and future perspectives. *Intractable & Rare Diseases Research*, 1(1):3–9.
17. MacArthur D. Orphan drugs in Asia-Pacific: From designation to pricing, funding and market access. *The Pharma Letter*, February 8, 2010.
18. Japan Intractable Diseases Information Center. What is an intractable disease? http://www.nanbyou.or.jp (access December 9, 2015).
19. Kanazawa I. Measures for dealing with "Intractable Diseases"? Past, present and future. *ICORD 2012 Conference*.
20. Simoens S. 2011. Pricing and reimbursement of orphan drugs: The need for more transparency. *Orphanet Journal of Rare Diseases*, 6:42.
21. Schuller Y, Hollak CE, Biegstraaten M. 2015. The quality of economic evaluations of ultraorphan drugs in Europe—A systematic review. *Orphanet Journal of Rare Diseases*, 10:92.
22. Desser AS, Gyrd-Hansen D, Olsen JA, Grepperud S, Kristiansen IS. 2010. Societal views on orphan drugs: Cross sectional survey of Norwegians aged 40 to 67. *British Medical Journal*, 341:c4715.

23. Sussex J, Rollet P, Garau M, Schmitt C, Kent A, Hutchings A. 2013. A pilot study of multicriteria decision analysis for valuing orphan medicines. *Value in Health*, 16(8):1163–1169.

24. Thornton P. 2010. *Opportunities in Orphan Drugs—Strategies for Developing Maximum Returns from Niche Indications.* London: Business Insights.

25. Côté A, Keating B. 2012. What is wrong with orphan drug policies? *Value in Health*, 15(8):1185–1191.

26. Winquist E, Bell CM, Clarke JTR, Evans G, Martin J, Sabharwal M, Gadhok A, Stevenson H, Coyle D. 2012. An evaluation framework for funding drugs for rare diseases. *Value in Health*, 15(6):982–986.

27. Meekings KN, Williams C, Arrowsmith J. 2012. Orphan drug development: An economically viable strategy for biopharma R&D. *Drug Discovery Today*, 17(13–14): 660–664.

28. Drummond M, Towse A. 2014. Orphan drugs policies: A suitable case for treatment. *European Journal of Health Economics*, 15:335–340.

29. Jarosławski S, Toumi M. Association between the prices of orphan drugs in oncology and the patient population sizes. PR1, *ISPOR 19th Annual European Congress*, Vienna, Austria, October, 2016.

11 Market Access of Vaccines in Developed Countries

11.1 INTRODUCTION

The term vaccine derives from the Latin *variola vaccinae*, a virus which was found to be able to prevent smallpox in humans when inoculated to healthy children, as demonstrated by Edward Jenner in 1796. Since then, various methods and technologies for vaccine development have been developed. Vaccination has dramatically reduced the incidence of several infectious diseases. The impact of vaccination has been estimated during the "vaccines for children era" (1994–2013) at the total number of more than 322 million prevented cases of infectious diseases, 21 million avoided hospitalizations, and 731,700 avoided deaths in the United States.[1]

Today, vaccines are required by payers, public health policy makers, and regulators to comply with the increasing standards of evidence-based prevention. It has also become harder to gain confidence from the society which becomes more concerned with vaccines' tolerability and safety.[2,3]

11.1.1 DEFINITION AND CLASSIFICATIONS

The World Health Organization (WHO) defines a vaccine as "a biological preparation intended to improve immunity toward a particular disease." The concept is based on stimulating the body's immune system versus an agent that is similar to the disease-causing pathogen or its components. The agent delivered by the vaccine is often either an attenuated or inactivated form of the pathogen or one of its components. The immune system will react by recognizing and destroying this agent while keeping a more or less sustainable memory of this reaction. This memory would make the elimination of the pathogen easier and more effective during future contact(s).[4]

Vaccines have different systems of classification that are based upon their components, the targeted pathogen, the mode of action, and their development technology. Currently, there is a clear distinction between two categories: (1) preventive vaccines (PVs) which are based on the traditional concept of preventing infectious diseases and (2) therapeutic vaccines (TVs) which represent a new concept of vaccinating to treat an already developed disease, and not necessarily an infectious one.

11.1.2 PREVENTIVE VACCINES

PVs are considered the second most important measure for reduction of human mortality after provision of safe drinking water.[5] PVs have become a powerful public health tool and almost all developed countries conceive and implement national or regional immunization programs for their populations. Improving health outcomes can have a positive impact on economic outcomes, for example, through longer working lives, higher productivity, improved educational outcomes, healthy and independent life expectancy (in relation to PVs for seniors), and reduced health care costs.[6]

According to 2013 The Organisation for Economic Co-operation and Development (OECD) health indicators, the national budgets in Europe allocated to health care represent an average of 9.0% of the gross domestic product (GDP), although on average only 3% of this budget is dedicated to prevention and less than 0.5% in vaccines.[3,7] The preventive health care budget is allocated to diverse areas including smoking cessation, reduction of alcoholism, improved nutrition, encouraging physical activity, screening and diagnostic tests, and higher uptake of vaccinations. New vaccines are awaited by physicians, payers, policy makers, and the global population. Notably, with the arrival of vaccines for all stages of life, from pediatric (diphtheria, tetanus, polio, etc.) to older population (e.g., pneumonia and herpes zoster vaccines), the policy makers speak of a life-course approach to vaccination. However, the occurrence of resistance to known pathogens and the appearance of new pathogen genotypes have raised concerns about the real effectiveness of PVs which progressively feature uncertainty. Also, the 2009 economic crisis has impacted the health care budgets in many countries with decreasing investments in prevention, as the benefits are not immediately perceivable.

Thus, the vaccines' market access (MA) has become a challenging discipline that requires specific expertise. Expert committees on vaccination have been established in developed countries with the purpose of examining and elaborating vaccines' recommendations and inclusion in national immunization programs, before staring funding and pricing process.

11.1.3 THERAPEUTIC VACCINES

The growing knowledge about the involvement of immune system in oncology has brought to light a new category of vaccines that qualify as therapeutic. This field is in development and only very few TVs exist on the market, for example, Provenge for the treatment of prostate cancer, approved in Europe and the United States markets, CVac for ovarian cancer, available in Dubai and CreaVax-RCC for the treatment of renal cell carcinoma, available in South Korea.[8]

TVs' development is characterized by high rates of failure in late phase-III clinical trials, caused mainly by the lack of efficacy. Unlike PVs, TVs typically follow pricing and reimbursement procedures similar to those for pharmaceuticals.

11.2 VACCINES' SPECIFICS

PVs are different from therapeutic drugs in many ways. The main features of this difference appear on the following three levels: (1) the target population is healthy and much larger than it is in the case for drugs, (2) the time of perception of benefit, and (3) the acceptance of adverse events which is higher for drugs that cure an already present disease than for vaccines which are administered to healthy people (Table 11.1).

Further, it takes as long as 12–24 months to produce a particular vaccine from the antigen to the finished product. Vaccines also feature further specifics with regard to their development, manufacturing, mode of action, targets, benefits, and MA as described below.

11.2.1 DEVELOPMENT

Vaccines' clinical development is lengthy. It takes around 10 years before delivering a single vaccine. As preventive medicines, PVs' development is based on larger scale clinical studies and closer monitoring than it is in the case of therapeutic drugs.

Vaccines' development is moreover risky. In fact, only 20% of preclinical candidates succeed to reach the market. The overall R&D cost for one successful candidate can range from $200 million up to $500 million, making it one of the most expensive and risky fields for investment.[9] Vaccines feature yearly investment that could be as low as $750 million, compared to 26.4 billion for pharmaceuticals.[10]

11.2.2 SAFETY

PVs' safety is a sensitive issue since they are given to large populations of healthy people. In this regard, payers and the society have higher expectations toward PVs than drugs. For example, the observed cases of the Guillain–Barré syndrome (GBS) following the vaccination against swine flu virus or cases of complex regional pain syndrome (CRPS) and postural orthostatic tachycardia syndrome (POTS), observed post-HPV vaccination fuelled suspicion in the society, whereas scientific competent authorities concluded that there is no link between these symptoms and the vaccine use.[11] Overall, post-marketing surveillance and adverse events follow-up in large populations are routine considerations when introducing new vaccines. They may involve both the regulatory and the HTA bodies.

11.2.3 BENEFITS AND COST-EFFECTIVENESS

When it comes to funding an intervention, payers appear to have more inclination to pay for immediate or short-term relief of illness rather than pay for preventing a condition that has not developed yet and for which effective drugs or treatment procedures exist. Therefore, assessing and demonstrating vaccines' benefits that are both humanistic and economic are crucial to gaining payer's acceptance. Measuring the impact/effectiveness in real life is critical and strongly linked to the vaccine coverage rates in the population. Both regulatory and HTA agencies may request real-world evidence, rather than only that from clinical trials, that the vaccine delivers benefits.

11.2.3.1 Humanistic Benefits

The humanistic benefit of large-scale immunization appears through many examples. For example, the eradication of devastating smallpox in Europe, the 99% of decrease in bacterial meningitis in the United States since 1989, thanks to the *Haemophilus influenza* vaccine, the 33,000 lives that are yearly saved and the 14 million infections that are prevented, thanks to the full vaccination from birth to adolescence in the United States.

Part of the benefits of vaccination is achieved, thanks to herd immunity that it grants. Thanks to vaccines, not only vaccinated people are individually protected, but importantly, the spread of disease to nonvaccinated individuals is prevented. At high rates of vaccination, the society receives large health benefits. Vaccinations have benefits throughout individuals' lives, because benefits of some vaccines are long-lasting and because there are vaccines for all stages of life.

11.2.3.2 Economic Benefit

Vaccines are associated with the economic benefits that are granted to the society upon the prevention of a certain disease and the avoidance of its treatment-related costs. Influenza vaccines are an example that generates up to $25 savings per person compared to corresponding treatment with oseltamivir.[12]

However, economic evaluations do not usually take into consideration the lost opportunity for economic growth or savings that can be achieved if broader diseases' complications and comorbidities are prevented or if resource allocation within the health care system is improved. It is believed that if policy makers were to include the appropriate factors for avoiding disease altogether (the intangible benefits of health) in the calculation, the value currently attributed to vaccines would be seen to be underestimated by a factor between 10 and 100.[13]

TABLE 11.1

Differences in Characteristics and Perceptions of PVs and Therapeutic Drugs

	Preventive Vaccines	Therapeutic Drugs
Action	Prevent	Heal/reduce risk
Population	Healthy population	Sick population or at risk
Population size	Typically entire healthy population concerned[a]	Target population depends on disease prevalence
Produced benefit	Individual and community level (herd immunity)	Individual level
Time of benefit	Future	Present time
Externality	High	Moderate
Acceptance of adverse events	Low	Moderate

[a] Target population may be narrowed if the disease concerns only men or women.

11.2.3.3 Cost-Effectiveness

Vaccines' cost-effectiveness is complex to evaluate and sensitivity analyses are critical. Also, determining the value of the vaccines' benefit needs a model-based research that should take into account four challenging factors, such as generating quality adjusted life years (QALYs) in children, their preferences, the discounting values, and finally the perspective selection.[14]

Vaccines' specificities such as the herd immunity, age shift, and serotype replacement also need to be taken into account. Dynamic models seem to be the most adapted to capture all the above-mentioned features.[15]

11.2.4 Market Typology

Growing vaccine prices and vaccine coverage rates (sales volume) remain one of the major drivers of the increase in vaccine expenditure. In the Unites States alone, the cost of administering all universally recommended vaccines to one child was multiplied by four to five times during the past decade.[16] However, in the European Union (EU), for a given vaccine, prices dropped over the past years (e.g., flu and HPV vaccines). Further, the maximum lifetime cost of vaccination for a citizen fully compliant with the national vaccination calendars (NVCs) ranges from 443€ to 3395€ depending on the broadness of the NVC and organization in the country.[7] This represents relatively low investment over lifetime compared to the benefits procured. Illustratively, in 2001, the drugs' global market exceeded $300 billion whereas vaccine sales were only about $5 billion.[17]

In Europe, vaccine markets can be classified as either public or private or mixed. Public market is more predominant in Spain, Italy, the United Kingdom, Sweden, Austria, and the Netherlands. These markets are managed by corresponding health authorities at national, regional, and local levels. The public markets operate through public procurement (tenders) and vaccination is performed within public infrastructures such as vaccination centers.[18]

Private reimbursement markets exists in, for example, France and Germany. They involve pediatricians and general practitioners, wholesalers, and pharmacists. They depend on prescriptions (like for drugs) and can sometimes generate substantial value for vaccine manufacturers. Awareness campaigns on vaccination are often driven by manufacturers.

Some markets are mixed, depending on the type of vaccines (pediatric/adolescent/adults) as it has place in Sweden.

11.3 OVERVIEW OF VACCINES' MARKET ACCESS IN DEVELOPED COUNTRIES

Vaccines' MA is a long and complex process with public health, clinical, ethics, and budget considerations. The key stakeholder in each country is an advisory committee that provides recommendations to decision makers and funding stakeholders. These committees are referred to as National Immunization Technical Advisory Groups (NITAGs).

The NITAG assesses the need for a given vaccination by considering the epidemiology, burden of the disease, and social consequences of a new vaccine introduction. NITAGs rarely issue preferential recommendation for one particular vaccine; however, they state the public health objective of a given immunization program. Their recommendation is intended for policy makers (Ministry of Health [MoH]) who decide on the inclusion of the new vaccine in the national immunization program which results in the vaccine being administered for free to the target population.

11.3.1 Overview of NITAGs' Processes

Across developed countries, NITAGs' feature general heterogeneity and lack of clear structure. Bryson et al. researched NITAGs' operating policies and revealed the lack of publicly available information across European countries.[19] It appears the NITAGs' lack of structure and transparency can make the decision analysis framework more susceptible to political influence.

In this regard, the VENICE project (Vaccine European New Integrated Collaboration Effort, *venice.cineca.org*), which involves 27 EU member states plus Norway and Iceland has been conceived to offer a transparent landscape for the European national vaccination programs.

11.3.1.1 NITAGs' Members

The number of NITAGs' members varies greatly from one country to another. According to Ricciardi et al., Hungarian NITAG is composed of 12 members while in the US, they have counted up to 48 members.[20]

11.3.1.2 Operations

NITAGs carry out their operations independently, but may be supported by agencies and an ad hoc expert committee. This is the case for the American NITAG: Advisory Committee on Immunization Practices (ACIP), and the German NITAG: Standing Vaccination Committee (*Ständigen Impfkommission* [STIKO]), which are supported by Centers for Disease Control and Prevention and the Robert Koch Institute, respectively. All NITAGs ultimately report to MoH.

11.3.1.3 Decision Criteria

Vaccines' inclusion on national immunization programs depend on several criteria (Table 11.2).

Not all developed countries use clear decision-making framework. Indeed, while Germany and the United States rely on a standardized methodology (the grades of recommendation, assessment, development, and evaluation), Canada, Italy, Spain, and Switzerland's Committees each have different, clearly listed criteria. France, Sweden, and the United Kingdom state only some of the decision-making criteria without defining a clear decision analysis framework. Australia, Belgium, and Hungary appear to not have

TABLE 11.2
Decision Criteria for Vaccine Recommendations

Decision-Making Criteria	European Countries								Non-European Country
	CH	DE	ES	FR	IT	NL	SE	UK	US
Disease burden	✓	✓	✓	✓	✓	✓	✓	✓	✓
Efficacy and effectiveness	✓	✓	✓		✓	✓	✓	✓	✓
Safety	✓	✓	✓	✓	✓	✓	✓	✓	✓
Feasibility of program implementation	✓	✓		✓		✓			✓
Cost-effectiveness evaluation	✓	✓	✓		✓	✓	✓	✓	✓
Clinical trial results		✓			✓		✓	✓	✓
Equity in access to the vaccine	✓		✓		✓		✓		✓
Recommendations of other liaison organizations							✓		✓

Source: Adapted from Ricciardi, G.W. et al., *Vaccine*, 33, 3–11, 2015.

Note: European countries—CH: Switzerland, FR: France, DE: Germany, IT: Italy, NL: the Netherlands, ES: Spain, SE: Sweden, UK: the United Kingdom; Non-European countries—US: the United States.

any defined decision-making criteria, as far as investigated by Ricciardi et al.[20]

11.3.2 IMPLEMENTATION OF NITAGS' RECOMMENDATIONS

NITAGs' recommendations are followed by policy decision-makers. In Germany, the Federal Joint Committee (G-BA) adjusts the vaccination guidelines based on the recommendations issued by the STIKO, and the final decision is issued by the MoH based on the G-BA guidance.[21] In the United Kingdom, the JVCI develops vaccination strategies and issues recommendations for the Department of Health which decides on the implementation of these recommendations and negotiates the price via a national tender.[22] In Italy and Spain, the national vaccination plan and calendar are agreed on national level by the MoH and also on a regional level with each region deciding its vaccination plan implementation and funding.[23]

11.3.3 TIME TO MARKET

The time from registration of a new vaccine until its availability to the population is often lengthy in European countries. On average, it could take up to 6.4 years (95% CI: 5.7 to 7.1 years) after the marketing authorization for a vaccine to reach its target population.[22] This time is driven by the length of the recommendation phase at the national level, which conditions the broad and funded population's access to a given vaccine.

11.4 OVERVIEW OF VACCINES' MARKET ACCESS IN A SELECTION OF EUROPEAN COUNTRIES AND THE UNITED STATES

11.4.1 AUSTRIA

The Vaccine Expert Committee (*Impfausschuss des OSR*) of the High Sanitary Council, the highest medical advisory board within the MoH, gives an annual advice on the vaccination schedule. The federal MoH usually takes this advice and publishes the schedule which is followed by all nine federal states.

The vaccination costs are shared among the federal MoH, the nine federal states and the social security system (*Hauptverband der Versicherungsträger*). Figure 11.1 summarizes MA pathway in Austria.[24] In Austria, there is a particularity: only pediatric and adolescents' vaccines are funded—by law, adult vaccines cannot.

11.4.2 DENMARK

The submission of the vaccine's dossier is made to the Danish Health and Medicines Authority (DMA) which is responsible for the licensing and reimbursement decision.

The National Board of Health is the health care authority in Denmark which provides professional advice on health issues to the MoH, the regions, and other authorities.

The five regions through their regional councils decide of the vaccine's use. They are in charge of funding and reimbursement. The regional health authorities advise the DMA before they make any decision. The regions buy pharmaceuticals via public procurement.

There is no regulatory mechanism in the Danish health service requiring the use of health technology assessment (HTA)

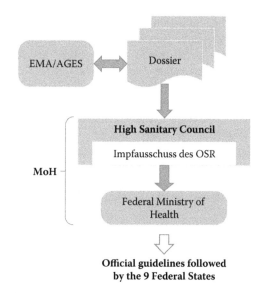

FIGURE 11.1 Vaccine's pathway for Market Access in Austria. AGES: Austrian Agency for Health and Food Safety, Impfausschuss des OSR: Immunization Advisory Committee, and EMA: European Medicines Agency.

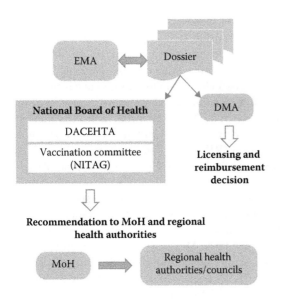

FIGURE 11.2 Vaccine's pathway for Market Access in Denmark. DACEHTA: Danish Centre for HTA, EMA: European Medicines Agency, MoH: Ministry of Health, and DMA: Danish Health and Medicines Authority.

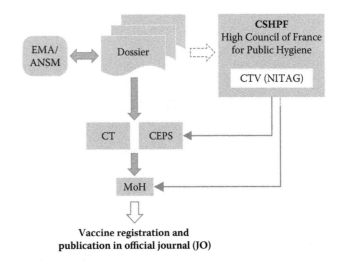

FIGURE 11.3 Vaccine's pathway for Market Access in France. ANSM: National Agency for Drug Security (*Agence nationale de sécurité du medicament*), CTV: Technical Committee of Vaccination, EMA: European Medicines Agency, CT: Transparency Committee, CEPS: Economic Committee of Health Products, and MoH: Ministry of Health.

in policy decisions, planning, or administrative procedures. At the national level, however, a number of comprehensive assessments of health technology have formed the basis for health policy decisions. Figure 11.2 summarizes MA pathway in Denmark.[24]

11.4.3 FRANCE

When the vaccine has obtained the marketing authorization (MAu), the company can submit a dossier to the Technical Committee on Vaccinations (CTV) in order to initiate the process of establishing guidelines on the vaccine's use.

Based on the CTV recommendation, a pricing and reimbursement (P&R) dossier is then submitted to the Transparency Committee (CT) for evaluation of medical value (HTA), and to the Economic Committee of Health Products (CEPS) for price negotiations.

The CTV provides an advice for recommendation to the MoH through the High Council of France for Public Hygiene (CSHPF).

Finally, the MoH gives the reimbursement and price decision, which may involve a request of additional elements, such as real-world evidence data. The inclusion of vaccine in the immunization schedule applies to the whole country. This system is being reformed in the frame of measures announced by the MoH in January 2016. A new Public Health Agency will be created and CTV will become a part of HAS.

Figure 11.3 summarizes MA pathway in France.[24]

11.4.4 GERMANY

The STIKO is the major federal commission concerned with vaccination issues (NITAG), appointed by the federal MoH.

Its recommendations are published by the Robert Koch Institute which is a central scientific agency that serves the federal MoH. The recommendations have no legal authority.

The G-BA makes the official recommendations 3 months after reviewing STIKO's report.

G-BA may also deviate from the recommendations of the STIKO.

Although the vaccination plan is nationally recommended, regions can decide independently to include in the schedule vaccinations for other diseases also, on the basis of the local epidemiological situation.

Licensed vaccines that are recommended by the STIKO are reimbursed by the statutory health insurance in all federal states of Germany. Licensed vaccines not recommended by the STIKO may be reimbursed by some health insurers voluntarily. Figure 11.4 summarizes MA pathway in Germany.[24]

11.4.5 ITALY

The National Vaccine Plan is prepared every 3 years by the MoH in collaboration with the National Institute of Health (*Istituto Superiore di Sanità* [ISS]), scientific societies, experts, and regional health authorities.

The price and reimbursement negotiations are made with the Pricing and Reimbursement Committee (CPR) and can include managed entry agreements if the vaccination program implies high-budget impact driven among others by large vaccine coverage rate (e.g., financial or outcome-based agreements).

Regional health authorities are responsible for the local implementation of the program, and can decide independently of the

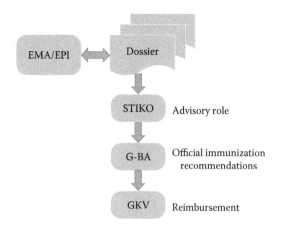

FIGURE 11.4 Vaccine's pathway for Market Access in Germany.[24] EMA: European Medicines Agency, EPI: Paul Ehrlich Institute, STIKO: Standing Vaccination Committee (*Ständigen Impfkommission*). G-BA: Federal Joint Committee (*Gemeinsamer Bundesausschuss*), and GKV: Statutory Health Insurance.

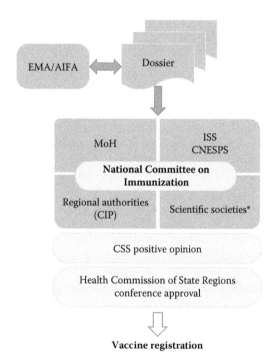

FIGURE 11.5 Vaccine's pathway for Market Access in Italy. CPR: Pricing and Reimbursement Committee, CSS: National Health Council, CIP: Interregional Prevention Commission, ISS: National Institute of Health (*Instituto Superiore di Sanita*), CNESPS: National Centre for Epidemiology Surveillance and Health Promotion. *For example, Center for HTA of the Università Cattolica del Sacro Cuore of Rome (Italy).

national plan, which vaccine to include according to the local epidemiological situation. Regions can also decide for additional vaccination policies at local level (e.g., varicella, boys HPV vaccinations). Figure 11.5 summarizes MA pathway in Italy.[24]

11.4.6 THE NETHERLANDS

The Dutch vaccine recommendation system is undergoing reforms as of 2016. The Ministry of Health, Welfare and

Sports (VWS) receives advice from Health Council (HC) before final decision.

HC is an independent scientific advisory committee which decides if a vaccine falls into the National Immunization Program (NIP) (100% funding), reimbursement (80%–100% funding), or individual purchase or private insurance (no funding). Criteria for HTA assessments have been published.[24] It receives advice from the Institute for Public Health and the Environment (RIVM) on epidemiology, and health economics, and from the National Healthcare Institute (ZIN).

Each province, plus the cities of Amsterdam and Rotterdam, has its own vaccine registry, which is responsible for vaccine distribution and the associated administration.

Another MA route is directly through the hospitals and clinical guidelines. Most vaccines for risk groups (e.g., influenza, zoster for rheumatics) follow this route. Reference pricing (RP) has not been used so far, but a new document of HC claims a role for RP-system in vaccines' assessment. Figure 11.6 summarizes MA pathway in the Netherlands.[24]

11.4.7 SPAIN

The Commission on public health chaired by the MoH has a technical working group (*Ponencia de vacuñas*) that is responsible for providing immunization recommendations to the nation, which then reach the Spanish Interterritorial Council (CISNS).

The CISNS is formed by National and Regional Ministries of Health (representatives of the 19 autonomous regions) and establishes the basic common health policy consensus for vaccination recommendations.

Then, the regional and local agencies are responsible for the allocation of funds and implementation of national health policies. Each autonomous community is responsible for purchasing vaccine for use in their region. Some communities provide additional vaccines that are not on the unique national recommended schedule (e.g., hepatitis A). Tenders are made directly with the manufacturer at the regional level. Figure 11.7 summarizes MA pathway in Spain.[24]

11.4.8 SWEDEN

Medical Products Agency (*Lakemedelsverket*) is responsible for licensing of vaccines.

The Public Health Agency (*Folkhälsomyndigheten* [PHA]) created in 2015 took the responsibility to make vaccine recommendation proposal for inclusion in the NIP. MoH takes decision on inclusion in NIP and funding.

In the case of reimbursed vaccines that are not a part of the NIP, a joint pricing and reimbursement application is submitted to the Dental and Pharmaceutical Benefits Board (*Tandvårds- och läkemedelsförmånsverket* [TLV]) for pricing approval.

Responsibility for pharmaceutical expenditure is held by the county councils, who select the product and purchase it. They may also impose additional restrictions on the usage

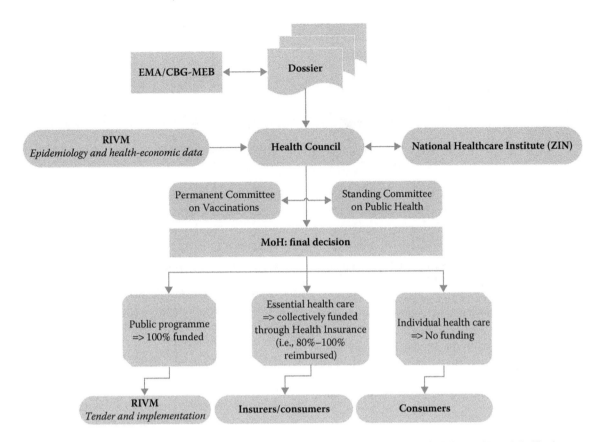

FIGURE 11.6 Vaccine's pathway for Market Access in the Netherlands. National Institute for Public Health and the Environment, VWS: The Ministry of Health, Welfare and Sports, NIP: National Immunization Program, and ZIN: National Healthcare Institute.

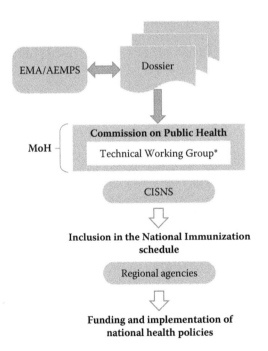

FIGURE 11.7 Vaccine's pathway for Market Access in Spain. AEMPS: Spanish Agency for Medicines and Health Products, CISNS: Spanish Interterritorial Council, and EMA: European Medicines Agency. * Ponencia de vacunas or Technical working group on vaccines includes groups, such as National Institute Carlos III, National centre for epidemiology, and so on.

of new drugs at the regional level. The county councils are responsible for the purchasing of vaccines and implementation of the NIP—they can regroup to issue regional tenders.

Figure 11.8 summarizes MA pathway in Sweden.[22]

11.4.9 THE UNITED KINGDOM

The vaccine has to be licensed first from the medicines regulatory authority and then the Public Health England (PHE) division, the Joint Committee on Vaccination and Immunisation (JCVI) makes a recommendation. JCVI may also ask the manufacturer to go through the centralized European Medicines Agency (EMA) procedure.

According to the 2009 National Health Service (NHS) Constitution for England, the Department of Health (DH) should ensure, so far as reasonably practicable, the implementation of JCVI recommendations.

JCVI is a standing advisory committee established under the NHS, the role of which is to advise the Secretary of State for Health and Welsh Ministers on matters relating to immunization. The JCVI's statutory functions do not relate to Scotland or Northern Ireland, although their Ministers may choose to accept its advice. JCVI makes a proposal every year without specific timelines and present their interest in having a specific treatment for a specific disease. The JCVI decisions are based on both science and economics as for the National

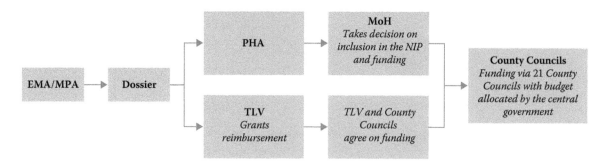

FIGURE 11.8 Vaccine's pathway for Market Access in Sweden. EMA: European Medicines Agency; Medical Products Agency (*Lakemedelsverket*); TLV: The Dental and Pharmaceutical Benefits Agency, and PHA: Public Health Agency (*Folkhälsomyndigheten*).

Institute for Health and Care Excellence (NICE), even though NICE does not address vaccines at all.[24]

Once the JCVI makes a recommendation, the DoH negotiates the price via national tenders and publishes the immunization recommendations in the Green Book.

To date, all four countries in the United Kingdom follow recommendations of the Green Book published by the DoH for immunization schedule.

Figure 11.9 summarizes MA pathway in the United Kingdom.

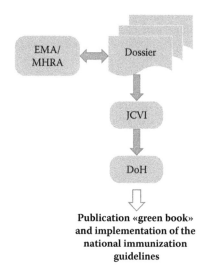

FIGURE 11.9 Vaccine's pathway for Market Access in the UK. MHRA: Medicines Health Regulatory Authority, EMA: European Medicines Agency, JCVI: Joint Committee on Vaccination and Immunisation, and DoH: Department of Health.

11.4.10 THE UNITED STATES

Vaccines are regulated by the Food and Drug Administration's (FDA's) Center for Biologics Evaluation and Research (CBER) to which vaccine developers must apply for permission to both develop and sell vaccines.

Once a vaccine is licensed for use, the Advisory Committee on Immunization Practices (ACIP) develops their recommendations which stand as public health advice. The committee provides guidance to the US Department of Health and Human Services (HHS) and the Centers for Disease Control and Prevention (CDC) on the control of vaccine-preventable diseases. The guidance is a written document on the vaccination features including the target population, the vaccine doses, and contraindication. This recommendation needs to be endorsed by the HHS and CDC to be enacted. Until now, all ACIP recommendations have been endorsed. These recommendations are the basis of the annual CDC immunization schedules for children and adults, and are used by health care providers in both public and private systems.[25]

For funding, recommended routine children vaccination is covered by the Vaccines for Children (VFC), which covers children who are eligible for Medicaid, uninsured, Native American, or underinsured. Children receive the vaccines free of charge. Also, according to the Affordable Health Care Act, health insurers are to provide the recommended vaccine free of charge to the policy holder, and insurers cannot charge premiums for vaccines. There are also health insurance plans that cover other population groups.

The process of vaccine funding in the United States is given in Figure 11.10.

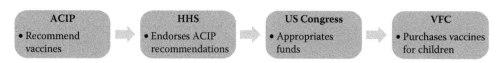

FIGURE 11.10 Funding for the VFC programs in the US. ACIP: Advisory Committee on Immunization Practices, HHS: Department of Health and Human Services, VFC: Vaccines for Children.

REFERENCES

1. Zhou, F., Singleton, J., Schuchat, A., Whitney, C.G. 2015. Benefits From Immunization During the Vaccines for Children Program Era—United States, 1994–2013. *Morb Mortal Wkly Rep* 63(16): 352–355.
2. Ricciardi, G.W. Market Access for Vaccines in Italy. [Paris]. 2014. Presented at European Market Access Diploma (EMAUD).
3. European Centre for Disease Prevention and Control. *Vaccine Hesitancy among Healthcare Workers and Their Patients in Europe—A Qualitative Study*. Stockholm: ECDC; 2015.
4. World Health Organization (WHO). Health Topics: Vaccines
5. Plotkin SL, Plotkin SA. A Short History of Vaccination. S Plotkin, W Orenstein and P Offit. Vaccines 5th edition. 2008. China, Saunders Elsevier.
6. Barnighausen, T. et al. 2014. Valuing Vaccination. *Proc. Natl. Acad. Sci. U.S.A* 111(34): 12313–12319.
7. Olivier Ethgen, Murielle Cornier, Emilie Chriv., Florence Baron-Papillon. 2016. The cost of vaccination through-out life: A western European overview, Human Vaccines & Immunotherapeutics. Accepted.
8. Boukhebza, H. et al. 2012. Therapeutic Vaccination to Treat Chronic Infectious Diseases: Current Clinical Developments Using MVA-Based Vaccines. *Hum. Vaccin. Immunother.* 8(12): 1746–1757, PM:22894957.
9. Andre, F.E. 2002. How the Research-Based Industry Approaches Vaccine Development and Establishes Priorities. *Dev. Biol. (Basel)* 110: 25–29, PM:12477303.
10. Berman, S., Giffin, R.B. 2004. Global Perspectives on Vaccine Financing. *Expert. Rev. Vaccines* 3(5): 557–562, PM:15485335.
11. HPV vaccines: EMA confirms evidence does not support that they cause CRPS or POTS. Available from: http://www.ema.europa.eu/docs/en_GB/document_library/Referrals_document/HPV_vaccines_20/European_Commission_final_decision/WC500196773.pdf.
12. Muennig, P.A., Khan, K. 2001. Cost-Effectiveness of Vaccination Versus Treatment of Influenza in Healthy Adolescents and Adults. *Clin. Infect. Dis.* 33(11): 1879–1885, PM:11692300.
13. Rappuoli, R., Miller, H.I., Falkow, S. 2002. Medicine. The Intangible Value of Vaccination. *Science* 297(5583): 937–939, PM:12169712.
14. Bo Standaert. Valuing Vaccines. 2014. Paris, GSK Vaccines. Presented at the European Market Access Diploma (EMAUD) Module 5.
15. Mondher Toumi. Medical Decision Analysis in Vaccines. 2014. Paris, Presented at the European Market Access Diploma (EMAUD) Module 5.
16. William Schaffner. Cost-Benefit of Vaccines: Considering Everyones Perspective. 2011. Healio Infectious Disease.
17. Thomas, P. The Economics of Vaccines. Harvard Medical International (HMI) World. 2002.
18. Christelle Saint Sardos. Vaccines Access in Europe. 2014.
19. Bryson, M. et al. 2010. A Global Look at National Immunization Technical Advisory Groups. *Vaccine* 28 Suppl 1: A13–A17, PM:20412990.
20. Ricciardi, G.W., Toumi, M., Weil-Olivier, C., Ruitenberg, E.J., Dankó, D., Duru, G., Picazo, J., Zöllner, Y., Poland, G., Drummond, M. Comparison of NITAG policies and working processes in selected developed countries. Vaccine. 2015 1;33(1):3–11.
21. Bryson, M. et al. 2010. A Global Look at National Immunization Technical Advisory Groups. *Vaccine* 28 (Suppl 1): A13–A17, PM:20412990.
22. Blank, P.R. et al. 2013. Population Access to New Vaccines in European Countries. *Vaccine* 31(27): 2862–2867, PM:23632307.
23. Christelle Saint Sardos. Vaccines Access in Europe. 2014.
24. Creativ-Ceutical Primary Research. 2013.
25. Centers for Disease Control and Prevention. Advisory Committee on Immunization Practices (ACIP). February 2, 2014.

12 France

12.1 STAKEHOLDERS

12.1.1 NAMES OF NATIONAL PRICING AND REIMBURSEMENT DECISION MAKERS

The Ministry of Social Affairs, the Ministry of Health, and the Ministry of Economy, Finance and Employment are the authorities in charge of the reimbursement legislation and decisions.[1,2] The Ministry of Health is the ultimate decision-making body in terms of reimbursement.

The Economic Committee for Health Care Products (*Comité Economique des Produits de Santé* [CEPS]) is a joint committee of ministers and health insurance funds. This body is in charge of setting the prices of reimbursed retail pharmaceuticals, some products in hospitals, and the reference prices (*tarif forfaitaires de responsabilité* [TFR]).[1,2,4]

The National Union of Health Insurance Funds (*Union nationale des caisses d'assurance maladie* [UNCAM]) is a third party payer (the head of health insurance funds). This body is in charge of setting the reimbursement rate (%) of pharmaceuticals.[1,2]

12.1.2 NAMES OF NATIONAL HEALTH TECHNOLOGY ASSESSMENT AGENCIES

The French National Authority for Health (*Haute Autorité de Santé* [HAS]) is an independent body that makes technical advice for including a pharmaceutical on the positive list but also recommendations and pharmacoeconomic evaluations. The following two HAS committees are involved in HTA[1,2]:

- The Transparency Committee (*Commission de la Transparence* [CT])
- The Economic and Public Health Assessment Committee (*Commission Evaluation Economique et de Santé Publique* [CEESP])

12.1.3 NAMES OF OTHER KEY STAKEHOLDERS (REGIONAL/LOCAL LEVEL)

- Regional Health Agencies (*Agences Régionales de Santé* [ARS]) are responsible for in-patient and out-patient organization of care for both public and private sectors[2]
- Public and private health establishments[1,2]

12.2 PRICING AND REIMBURSEMENT POLICIES

12.2.1 OVERVIEW OF THE SYSTEM

The marketing authorization is granted at the European level by the EMA (European Agency of Medicines) and at national level by French National Agency for Drug Safety (*Agence nationale de sécurité du médicament et des produits de santé* [ASNM]).* Following the marketing approval, manufacturers are required to submit a combined pricing and reimbursement application to both the CT and the CEPS. There are two positive reimbursement lists in France for drugs:

- List for drugs dispensed by retail pharmacies ("Liste des spécialités remboursables aux assurés sociaux").
- List for hospital drugs ("Liste des spécialités agréées aux collectivités"). This list is split into a list that contains drugs included in the diagnosis-related group (DRG) system (activity-based costing-T2A system) and a list that contains drugs for hospital out-patient use ("Liste de retrocession"). A product can be listed on the two lists above.
- Certain hospital-only costly drugs are charged to health insurance in addition to hospital stay fees based on DRG tariffs ("liste T2A").

Compulsory insurance health fund ("Assurance maladie") covers an estimated 90% of in-patient costs and two-thirds of the out-patient sector.

Assessment of drugs for reimbursement is carried out by the CT based on their clinical value, the so-called *Service Médical Rendu*/Actual Medical Benefit (SMR). SMR defines also reimbursement rate. There are five SMR levels, such as major, important, moderate, weak, and insufficient. These levels are based on the following five criteria:

- Severity of the disease and its impact on morbidity and mortality
- Clinical efficacy/effectiveness and safety of the medicine
- Aim of the drug: preventive, symptomatic or curative
- The therapeutic strategy as regards to therapeutic alternatives

---

* http://www.has-sante.fr/portail/upload/docs/application/pdf/2011-11/presentation_de_la_commission_de_la_transparence.pdf; http://www.sante.gouv.fr/IMG/ppt/Presentation_de_l_evaluation_des_medicaments_a_la_Haute_autorite_de_sante-2.ppt; http://www.has-sante.fr/portail/jcms/c_1267546/rapport-annuel-d-activite-2011; Local regulatory intelligence.

• Impact in terms of the public health (burden of disease, health impact at the community level, transposability of clinical trial results)

Prices of reimbursed medicines must be negotiated with the CEPS and are based on the level of ASMR (*Amélioration du Service Médical Rendu*/Added Actual Medical Benefit) appraised by the CT. Unless the product is first-in-the-class, the evaluation is done in comparison with existing therapies that are already enlisted.

There are five ASMR levels as follows:

• *ASMR I*: Major innovation/innovative product with substantial therapeutic benefit
• *ASMR II*: Important improvement in terms of therapeutic efficacy and/or reducing side effects
• *ASMR III*: Moderate improvement in terms of therapeutic efficacy and/or utility
• *ASMR IV*: Minor improvement in terms of therapeutic efficacy and/or reducing side effects
• *ASMR V*: No improvement over existing options but still recommended for reimbursement (e.g., generic drugs and me-to drugs)

HAS also determines the size of the target population.

HAS is currently aiming to introduce a new therapeutic index to assess medicines for pricing and reimbursement. The so-called ITR—or Index Thérapeutique Relatif—is designed to compare the benefit of medicines versus the appropriate comparator and would replace the current SMR/ASMR system.

At the end of the process, the Ministry of Health takes the decision for the registration of the medicines on the list of the reimbursed drugs, with a reimbursement rate and price. In theory, the opinion of the CT does not prejudge the decisions of the Ministry of Health in terms of reimbursement. However, in most cases, the Ministry of Health follows the decision of the CT. Decisions on pricing and reimbursement become legally applicable as soon as they are published by the Ministry of Health in the official gazette.

12.2.2   Reimbursement Process

The UNCAM fixes the reimbursement rate based on the SMR level granted by the CT. The registration on the list of reimbursed drugs is valid during 5 years. At the end of this period or at any time when significant new information is available, CT re-evaluates the SMR and ASMR levels.

There are four levels of reimbursement. They are as follows:

• 100% for medicines recognized as irreplaceable and especially expensive. In the case of chronic long-term illness, such as diabetes, severe arterial hypertension, HIV, and so on, medicines are reimbursed at 100% whatever the SMR level and the usual level of reimbursement may be.
• 65% for medicines with major or important SMR.
• 30% for medicines with moderate SMR.
• 15% for some medicines with weak SMR.

Drugs with insufficient SMR are not included on the list of reimbursed drugs.

12.2.3   Pricing Process

For reimbursed medicines, the price is regulated. Once a drug's eligibility for, and rate of, reimbursement has been decided, the price is negotiated between the manufacturer and the CEPS based on the therapeutic improvement (ASMR rating) of the drug granted by the CT on a scale of one (major therapeutic advance) to five (no improvement). The prices also depend on the following:

• The price of the drugs with the same therapeutic indication already available on the market
• The projected or observed sales' volumes
• The predictable or real conditions for use
• The size of the target population

Manufacturers are required to submit economic information on new pharmaceuticals to the CEPS drug pricing committee, listing benefits to patients, impact on overall expenditure, and any political considerations. Since October 2013, the cost-effectiveness of innovative medicines (ASMR I, II, or III) expected to have a significant impact on public finances (beyond 20 Million yearly turn over) is assessed by the CEESP.[1] Free pricing is applied for nonreimbursed medicines, which include the following:

• Over-the-counter products that are not on the reimbursement list
• Over-the-counter version of reimbursed products
• Products with mandatory prescription, but not reimbursed
• Hospital drugs included in the T2A list (DRG system) and freely negotiated between manufacturers and hospital pharmacies

Two categories of hospital medicines are subject to price negotiations with the CEPS as follows:

• Medicines that are listed in the T2A exclusion list (or non-T2A list)
• Medicines that are listed in the rétrocession list

Retail pharmacy-reimbursed drugs have controlled prices, but greater pricing freedom is granted for innovative drugs (ASMR I, II, and III) and ASMR IV products (with lower prices than comparators), for which pharmaceutical company can set the prices for reimbursement under some conditions. For ASMR I, II and III, a price should be in line with prices in Germany, Italy, Spain, and the United Kingdom, on the condition that they obtained a pharmacoeconomic advice of the CEESP.

12.3   TIME TO MARKET

For new medicines, the average time elapsed between marketing authorization and date of patient accessibility is equal to 316 days,[3] which is longer than elsewhere in the European Union. For generics, the average number of days until pricing and reimbursement approval is equal to 60 days.[4]

12.4 PRICE REGULATIONS

12.4.1 PRICING POLICY FOLLOWING THE MARKETING AUTHORIZATION

In France, pharmaceutical products cannot access the market before pricing and reimbursement approval.[1]

12.4.2 EXTERNAL REFERENCE PRICING

External reference pricing (ERP) is used as the main tool for the pricing of innovative medicines—that is medicines with ASMR ratings of I to III—as well as certain drugs which have obtained an ASMR rating of V compared to drugs which have recently been granted an ASMR rating of I, II, or III.

The CEPS and the LEEM pharmaceutical industry association agreed in the latest *accord-cadre* (2008–2012) (reaffirmed in the *accord-cadre* that was renewed in December 2012) not to set prices of concerned drugs below the lowest price among those charged in France's reference countries, such as Germany, Spain, Italy, and the United Kingdom.[5]

The agreement covers the five-year period which follows the reimbursement date, with the provision that high-cost drugs may be subject to further studies. The guarantee is expanded by one year for certain pediatric drugs and for drugs that are granted an indication extension with an ASMR of I to III, where the new indication is aimed at a different target population.

The European price guarantee might be shortened if a product qualifying for a European price obtains an extension of indication with an ASMR of IV or V for a population that is significantly larger than that for which the guarantee was initially granted.[5]

12.4.3 INTERNAL REFERENCE PRICING

The reference pricing system (*tarif forfaitaire de responsabilité* [TFR]) was introduced in September 2003, targeting off-patent drugs in therapeutic areas where generic penetration is low.

As a rule, new reference price groups are created for products with generic substitution rates below 50% (60% for high-turnover active ingredients) 12 months after genericization. Drugs with very low and/or nonprogressive substitution rates can be added to the TFR system after 24 or 36 months. There were a total of 337 groups as at the end of July 2012.[4] For these products, a reference price for reimbursement is set by the CEPS on the basis of the average price of generic versions of the off-patent molecule on the French market. The patient must pay out-of-pocket the difference between the reimbursement price and the full price of the drug dispensed, if the full price is higher.

12.4.4 PRICE CONTROL AT EX-FACTORY PRICE LEVEL

Prices are controlled for reimbursed pharmaceuticals. Free pricing is applied for all nonreimbursed pharmaceuticals in out-patient care, as well as for almost all products on the hospital positive list.[6,7]

12.4.5 PRICE CONTROL AT WHOLESALE LEVEL

Prices are controlled for reimbursed pharmaceuticals and a regressive mark-up scheme is applied for wholesalers.[7]

12.4.6 PRICE CONTROL AT PHARMACY RETAIL LEVEL

Prices are controlled for reimbursed pharmaceuticals via a regressive mark-up scheme, which can be reviewed by the Ministry of Health.[7]

12.4.7 MANDATORY PRICE REDUCTION ON BRAND PRICE AFTER GENERIC/BIOSIMILAR ENTRY

After generic entry, a price reduction equal to 20% is applied. A further price cut of 12.5% is implemented 18 months after the patent expiry of the originator drug if active ingredients are not included in the TFR. Mandatory price reduction is not applied after biosimilar entry.[8]

12.5 REIMBURSEMENT SPECIFICITIES

The positive list of reimbursed medicines is determined by the Ministry of Health after receiving technical advice from the CT. Only medicines that provide an improvement of medical service or savings in the cost of treatment are eligible for reimbursement. In France, there are two lists for reimbursed medicines, that is, positive list, as follows[4]:

- List for reimbursed medicines for out-patient care
- List of hospital medicines for hospital sector

The SMR defines the level of reimbursement as follows:

- *Important*: Drug is added to the reimbursement list and is reimbursed at 65%
- *Moderate*: Drug is added to the list and is reimbursed at 30%
- *Weak*: Drug is added to the list and is reimbursed at 10%–20%
- *No SMR rating/insufficient*: Drug is excluded from the positive reimbursement list

Once a reimbursed brand-name drug loses patent protection in France; any generic version that gains marketing approval is reimbursed at the same level as the originator.[4]

12.6 CHARACTERISTICS OF PUBLIC TENDERING

A tendering procedure exists for pharmaceuticals with alternatives purchased by hospitals (e.g., generics). In the public procurement procedure, favorable prices offered are an important criterion in the decision process.[4]

12.7 EXPENDITURE CONTROLS (SUPPLY SIDE)

12.7.1 DISCOUNTS/REBATES

Discounts offered by wholesalers to pharmacists on reimbursed drugs are subject to control. In addition, since January 2008, generics manufacturers have been prohibited from offering pharmacists extra rebates ("*marges arrières*") on top of the regular discounts. As of January 3, 2008, the maximum permitted discounts have been fixed as follows:

- 2.5% of the MSP for all branded drugs, except off-patent originals included in the TFR
- 17% of the MSP on generics (irrespective of whether or not they are included in the TFR), as well as off-patent branded originals included in the TFR[4,7]

12.7.2 CLAWBACK

All manufacturers of pharmaceuticals are concerned with the clawback system. If they engage in an agreement with the CEPS, they are not concerned by these taxes and pay a contribution, negotiated with the CEPS under the processes described in the "*accord-cadre*."[9]

12.7.3 PAYBACK

A payback clause ("*clause de sauvegarde*") between the CEPS and manufacturers requires manufacturers to cover any excess expenditure above the national target increase for reimbursed drug spending (known as the *taux k*) in a given year. The *taux k* is to be set at 0.4% in 2013, down from 0.5% in 2012 and 2011, and 1.0% in 2010. The clause applies to the following:

- Reimbursed retail drugs
- *Liste retrocession* drugs
- Since 2010, high-cost drugs excluded from the T2A hospital funding system

Repayments due from manufacturers are calculated as a percentage of their sales revenue on eligible drugs, and depend on the extent to which the *taux k* has been exceeded.[4]

12.7.4 PRICE-VOLUME AGREEMENTS

Price-volume agreements covering orphan drugs were set in the addendum to the CEPS-LEEM *Accord-Cadre* in October 2010.[4] For orphan drugs with an estimated cost per patient per year of more than €50,000:

- The price can be set at the European level (*taking into account* prices in Germany, Italy, Spain, and the United Kingdom), subject to a limit set by the CEPS

on total reimbursed turnover, based on the estimated number of patients.

- If sales exceed the pre-defined limit, the manufacturer is required to fund the provision of the drug to all eligible patients without restrictions.

12.7.5 OTHER MARKET ACCESS AGREEMENTS

Coverage with evidence (CED) agreements are being increasingly used in France (local regulatory intelligence).

12.7.6 PRICE FREEZES AND CUTS

Manufacturers who sign individual drug pricing conventions with CEPS, under the terms of the *accord-cadre*, are targeted by price cuts.[4]

Price cuts may be implemented because of the following reasons:

- There has been a substantial increase in sales volume since the initial reimbursement application was granted (e.g., following an indication extension).
- A comparable drug with similar efficacy has been marketed at a lower price.
- Pharmaceutical spending growth is likely to cause the annual target growth rate for healthcare (ONDAM) to be exceeded.

Manufacturers are also required to cut the ex-factory price of recently off-patent original drugs by 15% upon commercialization of the first corresponding generic. However, since March 2010, full or partial exemption from this requirement has been granted for products where the first generic version has been marketed more than 5 years after patent expiry.[4]

A further price cut is implemented 18 months after the patent expiry of the originator drug as follows, except for active ingredients included in the TFR:

- A 12.5% price cut at MSP on the off-patent branded original
- A 7% price cut at MSP on all corresponding generics

In addition, the ex-factory price of a generic drug is not permitted to exceed 40% of the pre-patent expiry MSP of the off-patent original drug.[4]

12.8 POLICIES TARGETED AT WHOLESALERS, PHARMACISTS, PHYSICIANS, AND PATIENTS

12.8.1 WHOLESALER AND PHARMACY MARK-UP

For reimbursed pharmaceuticals, wholesalers and pharmacists are remunerated through a regressive mark-up scheme. Both wholesalers' and pharmacists' margins are regulated.

Margins are free for nonreimbursed pharmaceuticals. The value-added tax (VAT) rate is 2.1% on reimbursed pharmaceuticals and 5.5% on nonreimbursed pharmaceuticals.[4]

12.8.2 GENERIC SUBSTITUTION

Physicians are not obliged by law to prescribe generics, but by agreement with health insurance funds, they have an indirect interest in prescribing generics. Physicians have a stake in the increase of generic prescriptions because the rise in the price of a consultation or a visit depends on the evolution of pharmaceutical expenditure (PE) under the convention signed with the health insurance funds.[4]

12.8.3 INN PRESCRIBING

Physicians are encouraged by agreement to write prescriptions by International nonproprietary name (INN) by agreement. In the most recent study carried out by the mutual funds, it appears that doctors prescribe by INN in the generics market at a rate of 12% (in terms of volume). The rate of generic prescription is growing significantly.[4]

12.8.4 PRESCRIPTION GUIDELINES

The HAS is required to develop guidelines for fully reimbursed diseases by health insurance fund. The guidelines are reviewed every three years. During this period, the list of reimbursed procedures (la liste des actes et prestations) is updated once a year.[4]

12.8.5 MONITORING OF PRESCRIBING BEHAVIOR

High-prescribing doctors (compared to other prescribers) can be identified and monitored by health insurances. For instance, health insurance can constrain the doctor's prescriptions to a "health insurance prior agreement" ("entente préalable").[4]

12.8.6 PHARMACEUTICAL BUDGETS DEFINED FOR PHYSICIANS

In France, there are no pharmaceutical budgets being applied for doctors or other health care providers, which means that there are no fixed prescribing budgets in terms of money available to health care professionals.[9]

The prescription volume or prescribing habits of general practitioners (GPs) and specialists are monitored by health insurance funds after consultation with the HAS. Through this, doctors are encouraged to prescribe the most economically viable pharmaceutical (generics) from several therapeutically similar alternatives.[9]

12.8.7 PRESCRIPTION QUOTAS

In France, prescription quotas are applied.*

12.8.8 FINANCIAL INCENTIVES FOR PHYSICIANS

In France, there is indirect financial incentive for doctors. If the national PE grows too fast, it may prevent an increase in consultation or visit fees for doctors at the national level.[4]

12.8.9 FINANCIAL INCENTIVES FOR PHARMACISTS

France aims to promote generics by creating a financial incentive for pharmacists for substituting a brand name pharmaceutical with a generic. For pharmacists, indirect financial incentives are applied; if they do not reach the recommended rate of substitution, the government can implement a discount reimbursed price on the brand.[9]

12.8.10 COPAYMENT FOR PATIENTS

Copayments comprise the difference between the retail price (100%) and the rate of reimbursement.[4]

Patients have to pay fixed co-payments as follows[4]:

- €0.5 per pharmaceutical pack purchased
- €1 per visit or consultation of the prescriber

REFERENCES

1. Lopes, S., Marty, C., Berdai, B. PHIS Pharma Profile France 2011. April; Pharmaceutical Health Information System; Commissioned by the European Commission, Executive Agency for Health and Consumers and the Austrian Federal Ministry of Health http://whocc.goeg.at/Literaturliste/Dokumente/CountryInformationReports/PHIS_Pharma%20Profile%20FR_2011_final.pdf.
2. Ispor Global Health care system roadmap, France, 2009. Available at: http://www.ispor.org/htaroadmaps/France.asp#2.
3. Efpia. Patient WAIT report, 2011.
4. Carone, G., Schwierz, C., Xavier, A. Cost-containment policies in public pharmaceutical spending in the EU. (European Economy. Economic Papers. 461. September 2012. Brussels. 62 pp). Available at: http://ec.europa.eu/economy_finance/publications/economic_paper/2012/pdf/ecp_461_en.pdf.
5. Leopold, C., Vogler, S., Mantel-Teeuwisse, A., de Joncheere, K., Leufkens, H.G.M., Laing, R. 2012. Differences in external price referencing in Europe. A descriptive overview. *Health Policy*, 104(1):50–60.
6. Pharmaceutical Pricing and Reimbursement Information (PPRI), France, 2008.

* http://www.ameli.fr/fileadmin/user_upload/documents/medecins-convention_version_consolidee_avenant7_01-06-2012.pdf.

7. Vogler, S. The impact of pharmaceutical pricing and reimbursement policies on generics uptake: Implementation of policy options on generics in 29 European countries-an overview. *Generics and Biosimilars Initiative Journal (GaBI Journal)*, 1(2): 93–100. Available at: http://whocc.goeg.at/Literaturliste/Dokumente/Articles/GJ2%2009j%20Vogler%20European%20focus%20generics.pdf

8. Mutualité Française-Report on generics drugs 2012-December 2012-http://www.medicamentsgeneriques.info/wp-content/uploads/2009/12/2012_12_M%C3%A9dicaments-g%C3%A9n%C3%A9riques.pdf

9. Espín, J., Rovira, J. Analysis of differences and commonalities in pricing and reimbursement systems in Europe. June; Andalusian School of Public Health; Commissioned by the European Commission, Directorate-General Enterprise.

FURTHER READING

Toumi, M. et al. 2015. Current process and future path for health economic assessment of pharmaceuticals in France. *Journal of Market Access & Health Policy*, 3. Available at: http://www.jmahp.net/index.php/jmahp/article/view/27902 (Accessed July 21, 2015).

13 Germany

13.1 STAKEHOLDERS

13.1.1 NAMES OF NATIONAL PRICING AND REIMBURSEMENT DECISION MAKERS

The Federal Joint Committee (*Gemeinsamer Bundesausschuss* [G-BA]) classifies pharmaceuticals into reference pricing groups, excludes pharmaceuticals from the Social Health Insurance (*Gesetzliche Krankenversicherung* [SHI]) benefit basket, and issues directives on pharmaceutical care.[1]

The National Association of Statutory Health Insurance Funds (*Gesetzliche Krankenversicherung-Spitzenverband* [GKV-SV]) is the central association of the health insurance funds at the federal level and is in charge of setting reference prices and prescribing targets.[2]

The Federal Ministry of Health (*Bundesministerium für Gesundheit* [BMG]) defines policies regarding health, including drafting of bills, ordinances, and administrative regulations.

13.1.2 NAMES OF NATIONAL HEALTH TECHNOLOGY ASSESSMENT AGENCIES

The Institute for Quality and Efficiency in Healthcare (*Institut für Qualität und Wirtschaftlichkeit im Gesundheitswesen* [IQWiG]) makes recommendations for the inclusion or exclusion of technologies and pharmaceuticals into the SHI benefit basket.[1] The IQWiG does not have any decision-making powers.[1,3]

The German Agency of Health Technology Assessment (*Deutsche Agentur für Health Technology Assessment* [DAHTA]) is part of the German Institute of Medical Documentation and Information (*Deutsches Institut für Medizinische Dokumentation und Information* [DIMDI]) and publishes health technology assessment (HTA) reports that cover different topics, such as prevention, diagnosis, therapy, rehabilitation, nursing, and methods.[4] The IQWiG may commission DIMDI for HTA reports. However, the benefit assessment by the IQWiG differs from the HTAs of the DIMDI.[5]

13.1.3 NAMES OF OTHER KEY STAKEHOLDERS (REGIONAL/LOCAL LEVEL)

Regional physicians' associations and regional health insurance funds make decision about health care provision.

13.2 PRICING AND REIMBURSEMENT POLICIES

13.2.1 OVERVIEW OF THE SYSTEM

The German health care system is characterized by a predominance of statutory health insurance with multiple competing sickness funds and private and public mix of providers. The statutory health insurance can be complemented by a private health insurance.

The market authorization can be granted at the European level by European Medicines Agency (EMA) or at national level by the Paul–Ehrlich Institute for Blood Products, Sera, and Vaccines and the Federal Institute for Pharmaceuticals and Medical Devices (*Bundesinstitut für Arzneimittel und Medizinprodukte* [BfArM]) for all other pharmaceuticals. Those two bodies are the official national licensing bodies for pharmaceuticals, and at the same time they supervise the safety of pharmaceuticals and medical devices.

The introduction of a mandatory benefit assessment process in Germany since January 2011, with the Act of the Reform of the Market for Medicinal Products (*Arzneimittelmarktneuordnungsgesetz* [AMNOG]), radically changed the environment of market access.[6] Price controls for new innovative reimbursed drugs were introduced through the Act, which created new requirements for comparator-driven evidence in clinical trials and significantly altered the pricing process to include discount negotiations. Although prices for new prescription medicines can still be set by the manufacturers, their validity is only guaranteed up to 6 months after launch. After this period, reimbursed prices depend on the early benefit assessment (EBA).[5] However, the visible list prices remain unchanged and discounts to SHI are in principle confidential and apply also to private insurers who are entitled to the same discounts. Free pricing is still in use for nonreimbursed drugs and those that are exempted from EBA.

From January 2014, benefit assessment for pre-AMNOG drugs is eliminated; instead, price freeze and mandatory rebates will continue until 2015.

13.2.2 REIMBURSEMENT PROCESS

Once the market authorization has been granted, all pharmaceuticals are eligible for reimbursement through SHI. However, reimbursement can be restricted to specific patient subgroups or indications. Some categories are excluded from reimbursement, such as nonprescription drugs, lifestyle drugs, drugs for minor illnesses, or with unproven therapeutic benefit. Off-label drugs are only reimbursed under certain conditions, such as positive recommendation by the G-BA.[8]

13.2.3 PRICING PROCESS

With the introduction of the AMNOG law, manufacturers are free to set their prices only within the first 6–12 months after marketing authorization. Indeed, manufacturer must seek pricing through the EBA procedure conducted by G-BA, in which manufacturers must demonstrate an additional benefit of the newly approved pharmaceutical compared to an appropriate compactor. Indeed, during the EBA, the G-BA classifies the new drug according to categories based on the additional therapeutic benefit: major, considerable, minor, nonquantifiable, none, and lesser therapeutic benefit than the comparator.[7] In case of additional benefit, pharmaceuticals' prices are subjected to price negotiations between the GKV-SV and the pharmaceutical company according to the level of additional benefit, the real prices of the product in other European countries, and the annual treatment costs of comparable drugs.[8] In case of agreement, the discounted price will take effect from the first day of the 13th month after launch.[9] If negotiations fail during the post launch-first year, an arbitration panel consisting of representatives of the sickness funds, the pharmaceutical industry, and neutral members is called to determine the discounted price within 3 months. The arbitration panel considers the benefit dossier, prices of comparable drugs, and European reference prices.[8] If manufacturer or GKV-SV rejects the arbitration panel's proposed price, either party can request a cost-benefit assessment from the IQWiG.

When no additional benefit is proven, newly approved drug is assigned to a reference price group. If there is no suitable reference price group, a reimbursement price is agreed so that the annual treatment costs are no higher than those of the appropriate comparator.[8] If the benefit is less than that of the comparator, more studies may be required or the drug may be excluded from reimbursement.[8]

13.3 TIME TO MARKET

In Germany, there is no delay in access to new drugs upon marketing authorization for brand medicines and generics. Time to market for pharmaceuticals in Germany is immediate upon marketing authorization.[10]

13.4 PRICE REGULATIONS

13.4.1 PRICING POLICY FOLLOWING THE MARKETING AUTHORIZATION

Manufacturers are free to set the ex-factory price for 6 months after marketing authorization, after which the price is either negotiated (added benefit) or pharmaceuticals are assigned to a reference price group (no added benefit). With the price negotiation taking up to 6 months, drugs with additional benefit enjoy up to 12-month free pricing.[9]

13.4.2 EXTERNAL REFERENCE PRICING

In Germany, external reference pricing (ERP) is applied since 2011. The country basket of Germany is composed of 15 European countries, such as Austria, Belgium, Czech Republic, Denmark, Finland, France, Greece, Ireland, Italy, the Netherlands, Portugal, Spain, Slovakia, Sweden, and the United Kingdom.[11] ERP is used as supportive criteria for setting the reimbursement price of pharmaceuticals that demonstrate a clinical added value. ERP is used during the price negotiations between the GKV-SV and the pharmaceutical company and by the arbitration panel.[8] Manufacturers should provide the actual ex-factory prices of the drug marketed in the reference countries before the start of the price negotiation.

13.4.3 INTERNAL REFERENCE PRICING

When no additional benefit has been established for a newly approved pharmaceutical, it is allocated to a reference price group with pharmacologically and therapeutically comparable pharmaceuticals. The setting of a reference price for a pharmaceutical is a two-step procedure involving the G-BA and the GKV-SV. First, the G-BA decides to which medicinal product group a price reference can be made and defines groups of medicinal products based either on the same active ingredients, pharmacologically therapeutically comparable active ingredients, or therapeutically comparable action. Second, the GKV-SV sets the reference price for the medicinal products using criteria that are regulated by laws. The list of reference prices for medicinal products is updated quarterly and published on DIMDI's website.[12]

13.4.4 PRICE CONTROL AT EX-FACTORY PRICE LEVEL

Like in most European countries, in Germany, price controls at ex-factory level are applied for reimbursed medicines whereas nonreimbursed medicines are freely priced.[13]

13.4.5 PRICE CONTROL AT WHOLESALE LEVEL

In Germany, price controls at the wholesale level are applied for prescription-only medicines (POM), reimbursed generics, and reimbursed over-the-counter (OTC).[13]

13.4.6 PRICE CONTROL AT PHARMACY RETAIL LEVEL

In Germany, price controls at the pharmacy retail level are applied for POM, reimbursed generics, and reimbursed OTC.[13]

13.4.7 MANDATORY PRICE REDUCTION ON BRAND PRICE AFTER GENERIC/BIOSIMILAR ENTRY

Price reduction on brand price after generic entry is mandatory in Germany.[13] The impact of generic entry on brand price depends on the number of competing generics.

13.5 REIMBURSEMENT SPECIFICS

Negative lists are used in Germany as a cost containment tool.[13] Indeed, there are several negative lists drawn by the G-BA for nonreimbursed drugs, such as lifestyle drugs, drugs for minor diseases, and OTCs.

13.6 CHARACTERISTICS OF PUBLIC TENDERING

In Germany, public tendering is applied in hospitals and for ambulatory care.[13] Tenders concern mostly generics, biosimilars, and only some of the branded pharmaceuticals.[13]

13.6.1 EXPENDITURE CONTROLS DISCOUNTS/REBATES

Discounts and rebates are widely used in Germany as cost-containment tools.[13] Indeed, mandatory rebate of 16% was applied on nonreference-priced drugs from August 2010. The rebate was supposed to end at the end of 2013.[6] However, with the elimination of the benefit assessments for pre-AMNOG pharmaceuticals, the mandatory rebate continued until 2015. This rebate was reduced to 6% on January 1, 2014 and increased to 7% on April 1, 2014. This mandatory rebate is also applied during the 12 months of free-pricing after the pharmaceutical's marketing authorization.

Rebates are also used for new pharmaceuticals. Indeed, negotiations between the manufacturer and the GKV-SV address the level of country-wide rebate on manufacturer's price for pharmaceuticals with added benefits.[6]

Discounts are negotiated between individual sickness funds and pharmaceutical companies for a single drug or an assortment of drugs.[8] These discounts vary widely and can amount to about 20% of the original price.

13.6.2 CLAWBACK

Clawbacks are not applied in Germany as tools for expenditure controls.[13]

13.6.3 PAYBACK

Paybacks are not applied in Germany as tools for expenditure controls.[13]

13.6.4 PRICE-VOLUME AGREEMENTS

Prices and volumes are defined during negotiation between the manufacturer and the GKV-SV[13] but such agreements exist mostly at the regional level. Manufacturers have to submit the treatment costs, expected volumes for their own product and for relevant competitors prior to the negotiation.

13.6.5 OTHER MARKET ACCESS AGREEMENTS

Risk-sharing agreements, apart from the price-volume agreements, are limited in Germany. Only fifteen agreements have been identified in the literature in 2013.[14]

13.6.6 PRICE FREEZES AND CUTS

Price freezes are widely used in Germany. The AMNOG law imposed a three-year price freeze, from August 2010 until December 31, 2013, on all pharmaceutical prices.[8] The purpose of this price freeze was to prevent an increase in the prices with an increase in the mandatory rebate to 16% in August 2010.

13.7 POLICIES TARGETED AT WHOLESALERS, PHARMACISTS, PHYSICIANS, AND PATIENTS

13.7.1 WHOLESALER MARK-UP

Wholesaler mark-up is regressive in function of the price.[15] The wholesaler mark-ups on manufacturer's price are between 6% and 12% of the pharmacy purchase price, with an average of 5%.[13]

13.7.2 PHARMACY MARK-UP

Pharmacy mark-ups are applied for POMs and OTCs. Mark-ups consist of a fixed component per package plus a linear mark-up of 3% of the wholesaler's price for POMs and are regressive for reimbursed OTCs.[15,13]

13.7.3 GENERIC SUBSTITUTION

Generic substitution is mandatory in Germany since 2002.[13] Indeed, the *aut idem* substitution was introduced in 2002 and pharmacists can substitute drugs without the physician's consent, unless it is specified otherwise on the prescription.[16] Pharmacists are obliged to substitute with a cheaper drug, when possible.

13.7.4 INTERNATIONAL NONPROPRIETARY-NAME PRESCRIBING

In Germany, as in the majority of European countries, INN prescribing is only indicative.[13] The physicians are not obliged to prescribe the active ingredient instead of the brand name.

13.7.5 Prescription Guidelines

The G-BA issues treatment guidelines that are legally binding for physicians in the SHI system.[13] They contain the general principles related to prescribing drugs and restrictions for particular drugs.

13.7.6 Monitoring of Prescribing Behavior

Physician's prescription behavior is monitored through electronic prescriptions, as in many other European Union countries.[13]

13.7.7 Pharmaceutical Budgets Defined for Physicians

Pharmaceutical budgets for physicians were abolished in 2002 and replaced by target volumes.[17] Regional physicians' associations define volume targets for individual doctors on the basis of the previous year's drug volume.

13.7.8 Prescription Quotas

Individual prescription target was introduced for physicians in 2002.[6] These targets are volume targets based on previous year's drug volume prescription.[13,17]

13.7.9 Financial Incentives for Physicians

Physicians can receive financial rewards or penalties depending on their prescription behavior.[13] Physicians who exceed their prescription target are requested to review their prescription behavior and to produce an explanatory letter.

13.7.10 Financial Incentives for Pharmacists

There are no financial incentives for pharmacists in Germany.[13]

13.7.11 Copayment for Patients

In Germany, copayments are set to 10% of the drugs' price[13] but not more than 10 EUR. Moreover, the total copayment per year is capped at 2% of total income, and at 1% of total income for chronically ill. Drugs priced 30% below their reference price are exempted from copayments, and children below the age of 18 years are excluded from copayments as well.

REFERENCES

1. The Federal Joint Committee. Available from: http://www.english.g-ba.de/downloads/17-98-2804/2010-01-01-Faltblatt-GBA_engl.pdf (accessed September 26, 2016).
2. The National Association of Statutory Health Insurance Funds. Responsibility for Healthcare. November 2012. Available from: http://www.gkv-pitzenverband.de/media/dokumente/presse/publikationen/Imagebroschuere_GKV-Spitzenverband_Einzelseiten_Englisch_2012.pdf (accessed September 26, 2016).
3. Institute for Quality and Efficiency in Health Care. Responsibilities, working methods and aims of IQWiG. Available from: https://www.iqwig.de/download/IQWiG_informationflyer.pdf (accessed September 26, 2016).
4. German Institute of Medical Documentation and Information website. HTA at DMDI. Last modified: March 2014. Available from: http://www.dimdi.de/static/en/hta/basicinfo-hta.pdf (accessed September 26, 2016).
5. German Institute of Medical Documentation and Information. Available from: http://www.dimdi.de (accessed September 26, 2016).
6. Henschke C, Sundmacher L, Busse R. Structural changes in the German pharmaceutical market: Price setting mechanisms based on the early benefit evaluation. *Health Policy* 109(3)P263–P269. March 2013.
7. Gerber-Grote A. IQWIG – Early benefit assessment and health economic evaluation: Experiences and challenges under the new law since January 1st. ISPOR European Congress 2012.
8. Ognyanova D, Zentner A, Busse R. Pharmaceutical reform 2010 in Germany. *Eurohealth* 17(1):11–13. 2011.
9. Pirk O, Hind Bouslouk M, Fricke FU. Additional patient related benefits are the key to price negotiation in Germany— Practical experience with benefit dossiers and the assessment process- *ISPOR 15th Annual European Congress*. November 2012. Available from: http://www.ispor.org/congresses/Berlin1112/presentations/W17_All%20Slides.pdf (accessed September 26, 2016).
10. The European Federation of Pharmaceutical Industries and Associations. Patients W.A.I.T. Indicator (Patients Waiting to Access Innovative Therapies). 2011. Available from: http://www.efpia.eu/documents/33/64/Market-Access-Delays (accessed September 26, 2016).
11. Toumi M, Rémuzat C, Vataire AL, Urbinati D. External reference pricing of medicinal products: Simulation-based considerations for cross-country coordination. European Union—For the European Commission. 2014. Available from: http://ec.europa.eu/health/healthcare/docs/erp_reimbursement_medicinal_products_en.pdf (accessed September 26, 2016).
12. German Institute of Medical Documentation and Information website. Reference Pricing Lists. Available from: http://www.dimdi.de/static/de/amg/festbetraege-zuzahlung/index.htm (accessed September 26, 2016).
13. Carone G, Schwierz C, Xavier A. Cost-containment policies in public pharmaceutical spending in the EU. European Economy - Economic Papers 461. September 2012. Available from: http://ec.europa.eu/economy_finance/publications/economic_paper/2012/pdf/ecp_461_en.pdf (accessed September 26, 2016).
14. Ferrario A, Kanavos P. Managed entry agreements for pharmaceuticals: The European experience. April 2013.
15. Vogler S, Habl C, Bogut M, Voncina L. Comparing pharmaceutical pricing and reimbursement policies in Croatia to the European Union Member States. *Croat Med J* 15;52(2):183–197. April 2011.
16. Rosery H. Reimbursement of drugs in Germany: A road map for the approval process. *ISPOR 9th Annual European Congress*, Copenhagen, Denmark. October 2006.
17. Sturm H, Austvoll-Dahlgren A, Aaserud M, Oxman AD, Ramsay C, Vernby A, et al. Pharmaceutical policies: Effects of financial incentives for prescribers. *Cochrane Database Syst Rev* 2007;3:CD006731.

14 Italy

14.1 STAKEHOLDERS

The main actors in the pharmaceutical system in Italy are: the Italian Medicines Agency (*L'Agenzia Italiana del Farmaco* [AIFA]), the Ministry of Health, the Ministry of Economics, and the Italian Institute of Public Health and the Regions.

14.1.1 NAMES OF NATIONAL PRICING AND REIMBURSEMENT DECISION MAKERS

AIFA, under the supervision of the Ministry of Health, is responsible for marketing authorization, pharmacovigilance, as well as pricing and reimbursement of all pharmaceuticals. AIFA is composed of two main advisory commissions: the Technical-Scientific Committee (CTS) and the Pricing and Reimbursement Committee (CPR) which are described as follows:

- The CTS makes decision on the reimbursement by assessing the therapeutic value of a product. This committee makes a decision on national marketing authorization, positive list revisions, limitation on prescribing specialists, and place in therapy for each drug.
- The CPR assesses the manufacturer's application, conducts price negotiations with manufacturers, and collects information on drug consumption using data from the National Observatory on the Use of Pharmaceuticals (OSMED).

14.1.2 NAMES OF NATIONAL HEALTH TECHNOLOGY ASSESSMENT AGENCIES

In Italy, the health technology assessment (HTA) is becoming more and more important at national, regional, and local levels with different timings, methods, and regulations, though its influence in terms of drugs pricing and reimbursement is not transparent.

HTA was formally introduced in the 2006–2008 National Health Plan by the Ministry of Health. Since 2009, the National Drug Agency (AIFA) conducts HTAs in collaboration with other National Institutions (i.e., OSMED).

The National Agency for Regional Health Services (*Agenzia Nazionale per i Servizi Sanitari Regionali* [AGENAS])—a noneconomic public institution in charge of providing technical and operative support to develop and implement the Italian National Health Services—was appointed in 2007 to provide technical and operational support for the development of HTA programs in the regions, though the regions themselves are responsible for implementation.

At present, just 5 out of Italy's 21 regions and autonomous provinces have established structures to include HTAs in

their health care decision-making process, such as Veneto (*Centro Regionale Unico del Farmaco* [CRUF]), Emilia–Romagna (*Osservatorio regionale per l'innovazione* [ORI], *Gruppo regionale farmaci oncologici* [GreFO]), Lombardy, Piedmont, and Tuscany (*Ente di Supporto Tecnicno Amministrativo Regionale* [ESTAR]). The role of regional HTA is mainly to assess drugs and make recommendations on their use.[13] The HTA analysis includes assessment of the technology clinical efficacy and safety as well as its cost-effectiveness and budget impact with the final aim of evaluating its risk-benefit profile.

14.1.3 OTHER KEY STAKEHOLDERS AT NATIONAL LEVEL

The role of other national stakeholders in the pharmaceutical is briefly described in Table 14.1.

14.1.4 REGIONAL AND LOCAL STAKEHOLDERS

The state of Italy includes 21 regional authorities (19 regions and two autonomous provinces). In each region, there are local health unities called *Aziende Sanitarie Locali* (ASL). The ASL is the operational part of the National Health Service (*Servizio Sanitario Nazionale* [SSN]). The ASLs are responsible for the daily management of the health services and for the coordination between hospitals. They are also in charge of the delivery of primary care, including contracts with general practitioners and independent hospitals (Aziende Ospedaliere, University Hospitals and Private Hospitals), provisions of occupational health services, health education, disease prevention, pharmacies, family advice, child health, and information services.[1]

14.2 PRICING AND REIMBURSEMENT OF PHARMACEUTICALS IN ITALY

14.2.1 OVERVIEW OF THE SYSTEM

The Italian reimbursement system covers all relevant diseases across the whole country and provides universal pharmaceutical coverage to the whole population (Italian citizens and legal residents). Reimbursable pharmaceuticals are included in a positive list, named the National Pharmaceutical Formulary (PFN) which is updated every 6–12 months.[1]

Public pharmaceutical expenditure accounts for approximately 75% of pharmaceutical expenditure in the country.[1] The pricing and reimbursement (P&R) processes are interlinked because they are both under AIFA responsibility and also because both decisions are undertaken within the same procedure.[1]

The pricing and reimbursement process for generics and biosimilars does not differ, though it may be faster than for new pharmaceuticals. According to a law decree issued in 2013,[2] the

TABLE 14.1
The Role of National Stakeholders in Italy

Body	Role
Ministry of Health	Control of AIFA activities and cooperation in the elaboration of pharmaceutical policy and regulations
	Regulation of wholesale, pharmacy, and retail policies
	Authorization and control of OTC pharmaceuticals advertising
	Regulation and control of production, marketing, and utilization of narcotic and psychotropic pharmaceuticals
	Updating of national official pharmacopoeia and price of galenics
Ministry of Economics	Control of AIFA activities and cooperation in the elaboration of pharmaceutical policy and regulations
Italian Institute of Public Health	Collaboration in the assessment of biological pharmaceuticals registration dossiers
	Quality control of pharmaceuticals
	Evaluation of clinical trials
	Pharmacovigilance, pharmacoepidemiologic, and pharmaco-utilization studies
Regions	Applying national rules
	Monitoring pharmaceutical expenditure
	Organizing services provided in their territories, coordinating and monitoring the work of the local health units (ASL). Through the State–Regions Conference, the regions nominate some members of the main AIFA Committees (CTS, CPR, CRS, and Board of Auditors)

Source: Folino-Gallo et al., Pricing and reimbursement of pharmaceuticals in Italy, *Eur J Health Econ.*, 9(3), 305–310, 2008.

price for such drugs is automatically set, without negotiation, in the reimbursement reference class, providing the manufacturer does not apply for a new price level. The law also sets the discount that should apply to generics and biosimilars which is equal to 45%–75% for Class A drugs and 30%–50% for Class H drugs (drug class description is reported in Section 14.2.3).[4]

Prices of non-reimbursed pharmaceuticals are freely established, with some limitations, by pharmaceutical companies. Prices of over-the-counter (OTC) pharmaceuticals may be rebated by pharmacists; thus the actual price may differ (a) from the official maximum price and (b) across the country.[1]

14.2.2 THE PROCESS

When marketing authorization is granted either by the European Medicines Agency (EMA) or the Italian Medicine Agency AIFA, the company may apply for reimbursement on the National Pharmaceutical Formulary (PFN).

The AIFA, assisted by the CPR and the CTS, has the responsibility to manage the different steps of pricing and reimbursement process. The negotiation procedure is conducted

following criteria based on product therapeutic value, pharmacovigilance data, price in other European Union (EU) member states, price of similar products within the same pharmacotherapeutic group, internal market forecasts, number of potential patients, and therapeutic innovation. The prices are negotiated at ex-factory level and also define the pharmacy retail prices. There are five stages in the P&R process as follows[1]:

1. The manufacturer submits a request to the Pricing and Reimbursement Unit (PRU) of the AIFA. The PRU does preliminary evaluations before the CPR meeting.
2. The CTS expresses an opinion on reimbursement classification through an evaluation of the clinical-therapeutic value, declares the start of the procedure, and issues the writ for the negotiation process to the CPR.
3. The CPR provides an evaluation of the manufacturers' dossiers (requests, reimbursement application) and also considers consumption and expenditure data provided by the Italian National Observatory for Pharmaceutical. The CPR makes comparisons with prices in other EU states and with similar drugs on the national reimbursement list (*Prontuario Farmaceutico Nazionale* [PFN]). The manufacturer's selling price is generally set at the European average. Prices are set by the AIFA for an initial two-year period.
4. The outcome of the negotiation is submitted to the CTS and then to the AIFA Management Board for final decision and approval.
5. The procedure concludes with the publication of the decision in the Official Journal of the Italian Republic (*Gazzetta Ufficiale* [GU]).

14.2.3 REIMBURSEMENT CLASSES

There are three categories of reimbursement as follows[3]:

- *Class A*: Provides 100% reimbursement of drugs dispensed to the public through retail pharmacies
- *Class H*: Provides 100% reimbursement for products restricted to hospital use
- *Class C*: Includes all the drugs that are not reimbursed by the SSN, such as prescription-only medicines (POM) and OTC products

In 2012, the Balduzzi decree[4] set a new class of reimbursement: C nn, where "C" stands for "not reimbursed" and "nn" stands for "not negotiated." Essentially, as soon as a new drug receives the European Medicines Agency (EMA) marketing authorization, it may be sold on the Italian market while waiting for the AIFA evaluation. This class was an attempt to speed up the uptake of new drugs not yet evaluated for reimbursement.

14.3 TIME TO MARKET ACCESS FOR DRUGS

Time to Market Access for drugs in Italy may be very long. According to a recent European Federation of Pharmaceutical Industries and Associations (EFPIA) report, for new drugs, the average number of days elapsed between market authorization and pricing and reimbursement approval is on average 347 days.[5] For generic medicines, the average number of days elapsed between market authorization and price and reimbursement approval is on average 90 days.[6]

Such timelines are further delayed by evaluations performed at regional and local (ASL/hospital) levels. The CNN class was therefore introduced to speed up the availability of new drugs on the Italian market.

14.4 PRICE REGULATION

14.4.1 PRICING POLICY FOLLOWING THE MARKETING AUTHORIZATION

Essentially, as soon as a new drug receives the EMA marketing authorization, it may be sold on the Italian market while waiting for the AIFA evaluation.

14.4.2 EXTERNAL REFERENCE PRICING

Since April 15, 2011, as a measure of cost-containment, external reference pricing (ERP) is used to set the price of generic and off-patent originator drugs. The calculations for the new reference prices are made on the basis of referencing with ex-factory prices in Germany, Spain, the United Kingdom, and France but are intended to take into account all EU member states.[7]

14.4.3 INTERNAL REFERENCE PRICING

Interchangeable medicines are grouped often by the same active ingredient (ATC-5) or chemical subgroup (ATC-4). Within each group, a reference price is defined which can be the lowest price or the average of a set of medicines in each group. In Italy, medicines are clustered by three levels, such as ATC-5, ATC-4, and ATC-38.

Regarding off-patent drugs, the SSN reimburses the lowest price among the pharmaceuticals with equal composition of active ingredients and with the same pharmaceutical form, the same method of administration, the same number of units, and the same unit dosage.

14.4.4 PRICE CONTROL (AT EX-FACTORY, WHOLESALE, AND PHARMACY RETAIL)

In Italy, a process of price control is applied at ex-factory, at wholesaler, and at pharmacy retail price levels for reimbursed drugs.[8]

14.4.5 MANDATORY PRICE REDUCTION ON BRAND PRICE AFTER GENERIC/BIOSIMILAR ENTRY

In the past, there was no mandatory price reduction on brand price after generic or biosimilar entry, but generics were required to be priced at least 20% below the price level of the originator drug.[1] Nowadays, a law decree issued in 2013[4] on generics and biosimilars establishes that the price for those drugs is automatically set, without negotiation, in the reimbursement reference class, in case the manufacturer does not apply for a new price level. The price of generics and biosimilars is set to be at least 45%–75% of Class A drugs and 30%–50% of Class H drugs.

14.5 COST-CONTAINMENT POLICIES

14.5.1 PRESCRIPTION GUIDELINES

The CTS issues prescribing guidelines (note) detailing the circumstances under which certain products can be reimbursed. As of September 2016, there were 94 notes.[9]

14.5.2 DISCOUNTS/REBATES

In Italy, there are different types of discounts and rebates granted to public payers[14]:

- Reduction of prices (price control: i.e., ex-factory price or wholesale/retail prices)
- Refunds by pharmaceutical companies back to public payers depending on the sales volume of medicines
- Risk-sharing agreements
- Shared risk of a potential spending over a pre-defined target

14.5.3 PAYBACK

Italy introduced a "payback" scheme in 2007, giving manufacturers the option to ask AIFA to suspend price cuts on specific drugs in return for paying an equal amount in funds to regional health administration. If, at the year's end, the drug expenditure is greater than the statutory ceiling (or if periodical monitoring during the year indicates it is likely to be surpassed), then the industry, wholesaler, pharmacists, and regions have to refund the excess.

For hospital drugs, regions and manufacturers refund both 50% of excessive spending, while territorial drugs are fully paid by manufacturers and distributors.[10] Payback applies to the entire pharmaceutical budget. The spending cap is defined at 11.35% for the retail sector and 3.5% for the hospital sector.[6,11]

14.5.4 PRICE-VOLUME AGREEMENTS

Manufacturers have to refund money to the state if pre-agreed sales/volumes are exceeded. Refunds may be in the form of lowering reimbursed prices.[8]

14.5.5 OTHER MARKET ACCESS AGREEMENTS

In Italy, the use of Market Access agreements has increased. They are mainly negotiated and decided at national level with AIFA and primarily managed at regional level. In practice, many MAAs are a mix of a financial-based agreement and payment by performance and mostly involve oncology drugs.[10]

14.5.6 Monitoring of Prescribing Behavior

In Italy, electronic prescription monitoring and guidelines linked to electronic systems (which support decision making and give feedback to the physician) are ways to improve prescription behavior.[8] When the system is fully operational, e-prescriptions will be sent to the regional authorities and the treasury in order to facilitate control over rational prescribing and expenditure. From 2009, the Italian government has created a specific flow of information and an official monitoring of hospital drug consumption and expenditure. As a result, regional authorities are forced to send this record to the Ministry of Health.[8]

14.5.7 Public Tenders

Public tendering is in place at the Local Health Unit and in the hospital sector for manufactures in order to further contain drug expenditure.[8]

14.5.8 Generic Substitution

Article 15 in law 135/2012 obliges doctors to indicate in the prescription the active ingredient to be used with no reference to a specific brand. Physicians can either decide to indicate only the International nonproprietary name (INN) as well as the brand name of the drug. They are also free to indicate a specific brand, obliging the pharmacists to not substitute the product, provided that they expressly give reasons for such choice.[8]

14.5.9 Copayment for Patients

All Italian regions have instituted a copayment system called "ticket," corresponding to a payment of between €1 and €2 per single prescription or per prescribed pack. The system differs significantly from region to region. Low-income people and people >65 years (with an income <€36,151.98 per year) are exempted from having to pay for these tickets. In some regions, specific patients (i.e., diabetics) are exempted from paying a ticket fee for medications used to control diabetes.[1]

A variable copayment can also be paid by patients, amounting to the difference between the reference price of a generic drug and the higher price of an equivalent version of the drug, if they choose to buy it anyway or if the doctor does not allow for drug substitution in the prescription.[8] The aim of this system is to make patients more sensitive to the cost of the medicines.

14.6 POLICIES TARGETED AT WHOLESALERS, PHARMACISTS

14.6.1 Wholesaler Mark-Up

The margins of reimbursed pharmaceuticals for the wholesalers are fixed by law at 3%.[12]

14.6.2 Pharmacy Mark-Up

The margins of reimbursed pharmaceuticals for the pharmacist are fixed by law at 30.35% of the net pharmacy retail price.[1,14] The pharmacy mark-up for reimbursed products is linear (fixed by the SSN).

For generic drugs, 8% of the mark-up is subtracted by industry mark-up and redistributed between wholesaler and pharmacy.[13]

REFERENCES

1. Folino-Gallo, P., Montilla, S., Bruzzone, M., Martini, N. Pricing and reimbursement of pharmaceuticals in Italy. *Eur J Health Econ.* 2008 Aug;9(3):305–310.
2. Ministry of the Health. Decree 4 Apr 2013. Criteri di individuazione degli scaglioni per la negoziazione automatica dei generici e dei biosimilari.
3. Creativ-Ceutical, Paris, internal proprietary database.
4. Law n. 189 Year 2012. Conversione in legge, con modificazioni, del decreto-legge 13 settembre 2012, n. 158, recante disposizioni urgenti per promuovere lo sviluppo del Paese mediante un piu' alto livello di tutela della salute. (12G0212)
5. EFPIA. Patient WAIT report, 2011.
6. Carone, G., Schwierz, C., Xavier, A. Cost-containment policies in public pharmaceutical spending in the EU (European Economy. Economic Papers. 461. Sep 2012. Brussels. 62 pp). Available from: http://ec.europa.eu/economy_finance/publications/economic_paper/2012/pdf/ecp_461_en.pdf (accessed October 06, 2016).
7. Leopold, C., Vogler, S., Mantel-Teeuwisse, A., de Joncheere, K., Leufkens, H.G.M., Laing, R. Differences in external price referencing in Europe. A descriptive overview. *Health Policy.* 2012;104(1):50–60.
8. Vogler, S. The impact of pharmaceutical pricing and reimbursement policies on generics uptake: Implementation of policy options on generics in 29 European countries-an overview. *Generics and Biosimilars Initiative Journal (GaBI Journal).* 1(2):93–100.
9. AIFA. Available from: http://www.agenziafarmaco.gov.it/it/content/note-aifa (accessed October 06, 2016).
10. AIFA. Order (Determina) 30 Octr 2014. Ripiano dello sfondamento del tetto dell'11,35% della spesa farmaceutica territoriale 2013, ai sensi della legge n. 222/2007 e ss.mm.ii. (Determina n. 1238/2014).
11. AIFA. Order (Determina) 30 Oct 2014. Ripiano dello sfondamento del tetto del 3,5% della spesa farmaceutica ospedaliera 2013, ai sensi della legge n. 135/2012 e ss.mm.ii. (Determina n. 1239/2014).
12. Law n. 122 Jul 2010. Conversione in legge, con modificazioni, del decreto-legge 31 maggio 2010, n. 78, recante misure urgenti in materia di stabilizzazione finanziaria e di competività economica.
13. World Health Organization, 2015. Access to new medicines in Europe: Technical review of policy initiatives and opportunities for collaboration and research. Available from: http://www.agenas.it (accessed October 06, 2016).
14. Negoziazione e rimborsabilità. AIFA. Available at http://www.agenziafarmaco.gov.it/it/content/negoziazione-e-rimborsabilità (accessed October 06, 2016).

15 Spain

15.1 STAKEHOLDERS

15.1.1 NAMES OF NATIONAL PRICING AND REIMBURSEMENT DECISION MAKERS

The General Subdirectorate of Quality of Medicines and Health Products (*Subdirección General de Calidad de Medicamentos y Productos Sanitarios* [SGCMPS]), which is a part of the Directorate General of Basic Services Portfolio of the National Health System and Pharmacy (*Dirección General de Cartera Básica de Servicios del Sistema Nacional de Salud y Farmacia*), is in charge of assessment of a combined pricing and reimbursement application and of the reimbursement decision.[1] The SGCMPS is seated within the Ministry of Health, Social Services and Equality (*Ministerio de Sanidad, Servicios Sociales e Igualdad* [MSSSI]).[2]

The Interministerial Commission for Pricing of Medicinal Products (*Comisión Interministerial de Precios de los Medicamentos* [CIPM]) is responsible for making the final decision on a pharmaceutical's price and reimbursement/funding.

The Spanish Agency for Medicines and Health Products (*Agencia Española del Medicamentos y Productos Sanitarios* [AEMPS]) is the regulatory body in charge of quality and safety control. It can give recommendation during the pricing process.[3]

15.1.2 NAMES OF NATIONAL HEALTH TECHNOLOGY ASSESSMENT AGENCIES

The national health technology assessment (HTA) agency (*Agencia de Evaluacion de Tecnologias Sanitarias* [AETS]) provides assessments on health, social, ethical, organizational, and economic impacts of the new innovative techniques and pharmaceuticals.[4] The AETS coexists with several HTA agencies in the autonomous region. Such HTA agencies include the Catalonian Technology Assessment and Medical Research Agency (*Agencia d'Avaluacio de Tecnologia Medica*), the Andalusian Health Technologies Assessment Agency (*Agencia de Evaluacion de Tecnologias Sanitarias de Andalucia*), the Galician Health Technologies Assessment Agency (*Axencia de Avaliación de Tecnoloxías Sanitarias de Galicia*), the Basque Health Technologies Assessment Service (*Osasunerako Teknologien Ebaluaketa*), and the Madrid's Technologies Assessment Unit (*Unidad de Evaluación de Tecnologías Sanitarias*). Their functions, as those of the AETS, include HTA and helping decision-making process. Their assessments are not mandatory for decision makers.

15.1.3 NAMES OF OTHER KEY STAKEHOLDERS (REGIONAL/LOCAL LEVEL)

Health care is decentralized with 17 autonomous communities (*Comunidades Autónomas*) providing basic health care services. Each of the 17 autonomous regions has competences in public health, health care services planning, and full control over their budget.

15.2 PRICING AND REIMBURSEMENT POLICIES

15.2.1 OVERVIEW OF THE SYSTEM

The National Health System (*Sistema Nacional de la Salud* [SNS]) provides health care to the Spanish population and certain types of Spanish residents.[5] The SNS is supervised by the MSSSI. While the MSSSI focuses more on pharmacovigilance, product approvals, cost containment, and long-term policies, the 17 autonomous regions are responsible for the health care delivery and finance. The SNS is funded through general taxation, and the central government provides financial support to each region, based on population and demographic criteria.

Market approval of medicines is granted by the European Medicines Agency or the AEMPS at national level. Marketing authorization holders are requested to submit a combined pricing and reimbursement dossier on approval of the medicines. The submission is assessed by the SGCMPS, and a report is brought to the CIPM. The CIPM is chaired by the Ministry of Health and is composed of representatives from the MSSSI and the Minister of Industry and Finances.[6] Since 2012, with the Royal Decree 16/2012, two additional representatives were included in the CIPM: rotating members from the autonomous communities.[11] The CIPM is responsible for price negotiations with pharmaceutical manufacturers and makes the final decision.

15.2.2 REIMBURSEMENT PROCESS

Spain has product-specific reimbursement eligibility. The decision of reimbursement and level of reimbursement is made by the SGCMPS. Several criteria such as the disease's severity, the therapeutic value and efficacy of the product, the level of innovation, the price of the product and of other therapeutic alternatives, and the impact on the budget compared with corresponding products are taken into account when deciding whether to reimburse a drug. Most of the reimbursed medicines are reimbursed at between 40% and 60%, according to the patient's income. However, pharmaceuticals for chronic illnesses are reimbursed at 90% and hospital medicines at

100%. Non-prescription medicines are not reimbursed and are not subjected to pricing decisions.[6]

15.2.3 Pricing Process

The key pricing policy for reimbursed medicines is price negotiations between the marketing authorization holder and the CIPM, based on the SGCMPS's evaluation. Indeed, the regulator sets the ex-factory price after negotiations with the marketing authorization holder. Criteria used to negotiate the price include severity of indication, usefulness of the medicine, needs of patient groups, rationality of costs, existence of therapeutic options, and degree of innovation. Although Spanish legislation does not regulate external reference pricing, according to our research, prices in 16 European Union (EU) member states (MS) are also considered for initial pricing and later revisions.

In general, innovative medicines are priced in alignment with other EU MS, but budgetary impacts are thoroughly considered. Generics are priced at least 40% below prices of originators[6] (reference pricing system). Non-reimbursement medicines are freely priced; however, manufacturers must inform the MSSSI of the price of their non-reimbursed pharmaceuticals.[11] Statutory pricing is applied for generics and me-to products, included in a cluster of the reference price system.[6]

The pricing process is still evolving in Spain. Indeed, according to Royal Decrees 9/2011 and 16/2012, cost-effectiveness should soon be taken into account (unknown implementation date). However, manufacturers are submitting some economic evidence that are currently unlikely to significantly alter decisions. The requirement for cost-effectiveness analysis is still not well formalized in Spain. A more standardized value-based approach is also beginning to be implemented, with the AEMPS elaborating therapeutic positioning reports. These reports are public and focus on clinical evidence and aim to support pricing and reimbursement decisions by authorities. However, the understanding and application of the economic concepts across the various regional autonomous authorities is variable.[14]

15.3 TIME TO MARKET

Market entry is highly delayed in Spain for new drugs. Indeed, average time to market is 352 days for a new drug and 60 days for generics.[7,8]

15.4 PRICE REGULATIONS

15.4.1 Pricing Policy Following the Marketing Authorization

Manufacturer must submit a combined pricing and reimbursement application before market entry.[6]

15.4.2 External Reference Pricing

External price referencing is used as supportive information and is applied when no similar medicines are on the Spanish market.[9] Spain's country basket is composed of 16 EU MS.[9] Even if the prices in the other countries are one of the many criteria, they often highly impact the negotiated price.

15.4.3 Internal Reference Pricing

An internal reference pricing system is in place for reimbursed generics and off-patent pharmaceuticals. The reference pricing system is based on the active ingredient and route of administration. The reference price level for each group is calculated based on the lowest daily treatment cost.[10] Generics and off-patent pharmaceuticals with no substitutable generic must be priced at or below the reference price level.[11] With the Royal Decree 16/2012, patented galenic forms (innovative forms), declared as galenic innovations, are no longer excluded from the internal reference pricing system.[11] Hospital medicines are included in independent groups.

15.4.4 Price Control at Ex-Factory Price Level

The ex-factory price of reimbursed medicines is controlled through the CIPM.

15.4.5 Price Control at Wholesale Level

The price of all the pharmaceuticals, including generics, is controlled at wholesale level.[12]

15.4.6 Price Control at Pharmacy Retail Level

The price of all the pharmaceuticals, including generics, is controlled at pharmacy level.[12]

15.4.7 Mandatory Price Reduction on Brand Price after Generic/Biosimilar Entry

Reduction on brand price after generic entry is mandatory in Spain (40% reduction).[8] After a reference pricing group is created when a generic enters the market, manufacturers must adjust the price of their brand product to the reference price level.

15.5 REIMBURSEMENT SPECIFICITIES

Positive and negative lists are applied in the outpatient sector, and thus, they are not relevant for the inpatient sector.[13] However, hospital formulary is in itself a positive list defined locally by hospitals. In 2012, with the Royal Decree 16/2012, 417 medicines belonging to 19 drug classes were delisted from the reimbursement list, 90 of which were maintained for reimbursement for only certain severe or chronic indications.[10]

15.6 CHARACTERISTICS OF PUBLIC TENDERING

Public tendering procedures have been linked to the hospitals and regions. However, this trend is quickly evolving with the centralized purchasing platform created in 2011 and applied in 2012 for influenza vaccine and some adult vaccines.[14] The centralized platform will purchase selected pharmaceuticals through tenders at national level and the product winning the tender will be reimbursed by the autonomous communities, only for the autonomous communities that want you to join this centralized platform. Tenders are already in place for hemophilia drugs, erythropoietin, and selected immunosuppressant therapies.[14]

15.7 EXPENDITURE CONTROLS

15.7.1 DISCOUNTS/REBATES

Discounts are used as a cost-containment tool in Spain and are regulated by law.[8,14] Discounts are shared by pharmaceutical companies, wholesalers, and pharmacies.[14] A mandatory 7.5% discount is applied on the public prices of innovative drugs, patented retail, and hospital drugs since June 2010 (4% for orphan medicines), and a mandatory discount of 15% is applied on retail and hospital products marketed for more than 10 years with no generic version on the market since 2011 (Royal Decree 9/2011).

15.7.2 CLAWBACK

Clawback policies are applied to pharmacies and clawback's thresholds and percentages are regulated by Royal Decree 4/2010.[8]

15.7.3 PAYBACK

Paybacks are applied to manufacturers who are requested to pay back a percentage of the quarterly sales of reimbursed pharmaceuticals (from 1.5% to 2%).[8,14]

15.7.4 PRICE-VOLUME AGREEMENTS

Price-volume agreements are not used at national level but are sometimes used at regional level by the autonomous communities and hospitals for high-cost innovative drugs.[8]

15.7.5 OTHER MARKET ACCESS AGREEMENTS

Market access agreements are concluded mainly at regional level by the autonomous communities, but most of them are confidential. There are just a few examples at national level. These agreements are mostly performance-based agreements (payment for performance) for expensive drugs. There are no regulations or incentives at national level.[8,15]

15.7.6 PRICE FREEZES AND CUTS

Price cuts are applied as a cost-containment tool and are regulated by Royal Decree. Retail price cuts were imposed on generics by the Royal Decree 4/2010 (up to 30%) and on a selection of innovative drugs, based on the price in other EU MS in 2010.[8]

15.8 POLICIES TARGETED AT WHOLESALERS, PHARMACISTS, PHYSICIANS, AND PATIENTS

15.8.1 WHOLESALER MARK-UP

Wholesaler mark-ups are fixed by law and are regressive in Spain.[8]

15.8.2 PHARMACY MARK-UP

Pharmacy mark-ups are fixed by law and are regressive in Spain.[8]

15.8.3 GENERIC SUBSTITUTION

Generic substitution is mandatory, and pharmacists have the obligation to dispense the lowest-price product (Royal Decree 16/2012).[8,10]

15.8.4 INTERNATIONAL NONPROPRIETARY NAME (INN) PRESCRIBING

The INN prescribing became mandatory with the Royal Decree 16/2012. Physicians are requested to prescribe per active ingredient unless there is a specific medical need for the brand product or there is no authorized generic.[10]

15.8.5 PRESCRIPTION GUIDELINES

The MSSSI is in charge of proposing and implementing guidelines on health policy, clinical practice, and health care. Clinical practice guidelines (*Guías de Práctica Clínica*) set recommendations on the diagnosis and treatment of diseases and are indicative.[16] The MSSSI also launched an evidence-based clinical knowledge system for physicians at national level in early 2014.

15.8.6 MONITORING OF PRESCRIBING BEHAVIOR

Prescriptions are monitored at national and regional levels.[8] "Prior inspection patient visas," that is, authorizations (*visados previos de inspección*), are used at national level to ensure that expensive and specific pharmaceuticals are used for their intended purpose.[17] The visa system is also used at regional level by some of the autonomous communities such as in Andalucía, which uses visas for atypical anti-psychotics. Moreover, electronic prescribing systems are used at regional level to monitor prescribing behavior.[18]

15.8.7 PHARMACEUTICAL BUDGETS DEFINED FOR PHYSICIANS

Some of the autonomous communities define target budgets for physicians.[8] Target budgets vary from one community to another.

15.8.8 Prescription Quotas

Prescription quotas and percentage of generic prescribing are also applied at regional level for physicians.[8]

15.8.9 Financial Incentives for Physicians

Target-based incentives are used by regions to promote good prescription behavior.[8]

15.8.10 Financial Incentives for Pharmacists

There are no financial incentives for pharmacists in Spain.[8]

15.8.11 Copayment for Patients

Copayment became linked to patient's income (starting from €18,000/year) since 2012 with the Royal Decree 16/2012.[7,8] Copayments range between 40% and 60% for active workers and between 10% and 60% for retired workers.[19]

REFERENCES

1. Directorate General of Pharmacy and Health Products Organisation: The Ministry of Health, Social Services and Equality. Available from: http://www.msssi.gob.es/organizacion/ministerio/organizacion/sgralsanidad/dgfarmayps.htm and https://www.msssi.gob.es/profesionales/farmacia/organizacion.htm (accessed October 06, 2016).
2. The Ministry of Health, Social Services and Equality. Available from: http://www.msssi.gob.es/en/organizacion/ministerio/home.htm (accessed October 06, 2016).
3. The Spanish Agency for Medicines and Health Products. Available from: http://www.aemps.gob.es/en/home.htm (accessed October 06, 2016).
4. The national Health technology assessment agency. Available from: http://www.isciii.es/ISCIII/es/contenidos/fd-el-instituto/fd-organizacion/fd-estructura-directiva/fd-subdireccion-general-investigacion-terapia-celular-medicina-regenerativa/fd-centros-unidades/agencia-evaluacion-tecnologias-sanitarias.shtml (accessed October 06, 2016).
5. The Ministry of Health, Social Services and Equality. National Health system in Spain. 2012. Available from: https://www.msssi.gob.es/en/organizacion/sns/docs/sns2012/SNS012—Ingles.pdf (accessed October 06, 2016).
6. Martínez Vallejo M, Ferré de la Peña P, Guilló Izquierdo MJ, Lens Cabrera C. PHIS Pharma Profile Spain. 2010. Available from: http://whocc.goeg.at/Literaturliste/Dokumente/CountryInformationReports/Spain_PHIS_PharmaProfile_2010.pdf (accessed October 06, 2016).
7. The European Federation of Pharmaceutical Industries and Associations. Patients W.A.I.T. Indicator (Patients Waiting to Access Innovative Therapies). 2011. Available from: http://www.efpia.eu/documents/33/64/Market-Access-Delays (accessed October 06, 2016).
8. Carone G, Schwierz C, Xavier A. Cost-containment policies in public pharmaceutical spending in the EU. European Economy—Economic Papers 461. September 2012. Available from: http://ec.europa.eu/economy_finance/publications/economic_paper/2012/pdf/ecp_461_en.pdf (accessed October 06, 2016).
9. Toumi M, Rémuzat C, Vataire AL, Urbinati D. External reference pricing of medicinal products: Simulation-based considerations for cross-country coordination. European Union - For the European Commission. 2014. Available from: http://ec.europa.eu/health/healthcare/docs/erp_reimbursement_medicinal_products_en.pdf (accessed October 06, 2016).
10. Dylst P, Simoens S, Vulto AG. Reference pricing systems in Europe: Characteristics and consequences. *Generics and Biosimilars Initiative Journal (GaBI Journal)*, 1(3–4): 127–131. 2012. Available from: http://gabi-journal.net/reference-pricing-systems-in-europe-characteristics-and-consequences.html#R13 (accessed October 06, 2016).
11. Lee JL, Fischer MA, Shrank WH, Polinski JM, Choudhry NK. A systematic review of reference pricing: Implications for US prescription drug spending. *Am J Manag Care* 18(11):e429–e437. 2012.
12. Vogler S. The impact of pharmaceutical pricing and reimbursement policies on generics uptake: Implementation of policy options on generics in 29 European countries-an overview. *Generics and Biosimilars Initiative Journal (GaBI Journal)*, 1(2):93–100. 2012. Available from: http://gabi-journal.net/the-impact-of-pharmaceutical-pricing-and-reimbursement-policies-on-generics-uptake-implementation-of-policy-options-on-generics-in-29-european-countries%E2%94%80an-overview.html (accessed October 06, 2016).
13. Vogler S, Habl C, Bogut M, Voncina L. Comparing pharmaceutical pricing and reimbursement policies in Croatia to the European Union Member States. *Croat Med J* 15;52(2): 183–197. 2011.
14. Vogler S, Zimmermann N, Habl C, Piessnegger J, Bucsics A. Discounts and rebates granted to public payers for medicines in European countries. *Southern Med Review*, 5(1):38–46. 2012.
15. Ferrario A, Kanavos P. Managed entry agreements for pharmaceuticals: The European experience. April 2013. EMiNet, Brussels, Belgium.
16. Clinical Practice Guidelines. Library of clinical practice guidelines of the National Health System. 2014. Available from: http://portal.guiasalud.es/web/guest/guias-practica-clinica (accessed October 06, 2016).
17. Schoonveld E. *The Price of Global Health: Drug Pricing Strategies to Balance Patient Access and the Funding of Innovation.* Gower Publishing, Surrey, England, 2011.
18. Medinilla Corbellini A, Giest S, Artmann J, Heywood J, Dumortier J. Country Brief: Spain. eHealth Strategies Report. October 2010. Available from: http://ehealth-strategies.eu/database/documents/spain_countrybrief_ehstrategies.pdf (accessed October 06, 2016).
19. Pharmaceutical Pricing and Reimbursement information. Recent changes in pharmaceutical policy measures and developments in pharmaceutical expenditure. 2013. Available from: http://whocc.goeg.at/Literaturliste/Dokumente/CountryInformationPosters/Spain_PPRI%202013.pdf (accessed October 06, 2016).

16 Sweden

16.1 STAKEHOLDERS

16.1.1 NATIONAL PRICING AND REIMBURSEMENT DECISION MAKERS

In Sweden, the key organization involved in the pricing and reimbursement process for drugs is the Dental and Pharmaceutical Benefits Agency (*Tandvårds-och läkemedelsförmånsverket* [TLV]), formerly known as the Pharmaceutical Benefits Board (*Läkemedelsförmånsnämnden* [LFN]). The TLV is a central government agency that determines whether a pharmaceutical product or dental care procedure shall be subsidized by the state and has decisive power at a national level.[1] It makes a simultaneous decision on the drug's pricing and reimbursement.

16.1.2 NATIONAL HEALTH TECHNOLOGY ASSESSMENT AGENCIES

The TLV is the main health technology assessment (HTA) agency that evaluates all retail drugs and directly influences their inclusion in the reimbursement system, the Pharmaceutical Benefit Scheme.

Another Swedish HTA organization is the Swedish Agency for Health Technology Assessment and Assessment of Social Services (*Statens beredning för medicinsk och social utvärdering* [SBU]), founded in 1987.[2] The SBU assesses health care technologies from medical, economic, ethical, and social standpoints. The SBU also produces guidelines and reports and disseminates information on new drugs and technologies. The assessments of SBU are based on systematic literature reviews of published research; the manufacturers cannot make submissions.

Although the SBU performs HTA analysis, it does not directly influence pricing and reimbursement decisions in Sweden. Instead, SBU's independent and scientifically based assessments are sources of knowledge for decision-making authorities such as the National Board of Health and Welfare, the Medical Products Agency (MPA), and the TLV, as well as for health care providers and health centers, when developing standard of care guidelines. The SBU's HTAs are not made for new drugs entering the market but, in general, are focused on evaluation of management of a disease. Suggestions for topics for assessment by SBU may come from various sources, such as individuals, organizations, government authorities, and the SBU Scientific Advisory Committee.

16.1.3 OTHER KEY STAKEHOLDERS

The state, through the Ministry of Health and Social Affairs, is responsible for overall health care policy. Other key stakeholders directly involved in health care, in addition to the TLV and SBU presented above, include the Medical Products Agency, the National Board of Health and Welfare, the Ministry of Health and Social Affairs, and the county councils.

The Medical Products Agency (*Läkemedelsverket* [MPA])[3] is the Swedish national authority responsible for the regulation and surveillance of the development, manufacturing, and sale of pharmaceuticals and other medicinal products. A major part of the MPA's work is devoted to the approval of medicines. The marketing authorization for a new drug is granted by the European Medicines Agency (EMA) and/or the MPA at national level. The MPA is not involved in the pricing and reimbursement process of a drug or the use of the drug in practice. However, it produces recommendations for medical treatment in various therapeutic areas.

The National Board of Health and Welfare (Socialstyrelsen)[4] is a government agency with a very wide range of activities and different duties within the fields of social services, health and medical services, patient safety, and epidemiology. It gives support and exerts influence through data collection, analysis, and dissemination of information; development of standards based on legislation and data; and the maintenance of health data registers and official statistics. While the overall responsibility of health care rests with the Ministry of Health and Social Affairs, supervision of the delivery of health care is performed by the National Board of Health and Welfare.

The Ministry of Health and Social Affairs (Socialdepartementet)[5] is a regulatory body to which the TLV, MPA, and the National Board of Health and Welfare report. It has an overall planning and legislative authority.

Lastly, pursuant to the Health and Medical Services Act of 1982, the health care system is decentralized and the county councils[6] are responsible for the funding and provision of health care services to their populations. County councils are political bodies elected by the public every 4 years. They are responsible for the local health care budget and have the authority to recommend a drug on local formularies.

16.2 PRICING AND REIMBURSEMENT POLICIES

16.2.1 OVERVIEW OF THE SYSTEM

The Swedish health care system is a national health service, financed mainly through proportional taxes levied by county councils and municipalities and the rest through state subsidies and user charges.[7]

It is a highly decentralized system. Although health care goals and policies are decided at the national level, the provision of care is determined by county councils and, in some cases, by municipalities; both have considerable flexibility in deciding how to plan and deliver health care.

Thus, although the Ministry of Health and Social Affairs is responsible for health care policy and funding, county councils cover majority of the operations and costs of hospitals, primary care centers, and both inpatient and outpatient drugs. Long-term care for the mentally ill and elderly is carried out by municipalities.

16.2.2   REIMBURSEMENT PROCESS

Retail pharmaceuticals may be included in the Pharmaceutical Benefits Scheme, provided that the pharmaceutical company's application gains approval from the TLV.[8] Medicines for inpatient care are not covered by the Pharmaceutical Benefits Scheme but are funded by the hospital budgets.

In Sweden, there is a combined pricing and reimbursement (P&R) application process and decision. In the same manner, non-reimbursed drugs, that is, non-reimbursed prescription-only (POM) medicines and most over-the-counter (OTC) medicines, may be freely priced. The TLV's Pharmaceutical Benefits Board meets once a month and makes all P&R decisions.

The process starts when a manufacturer submits an application to TLV. The following must be enclosed for it to be deemed complete: a proposed price, general product information, clinical evidences, health economic analysis, information on the subpopulation of interest, treatment landscape (alternatives and comparators), and estimated costs and duration of treatment. Documents are evaluated by the expert committee consisting of a health economist, a clinical expert, and a legal advisor. Then, the TLV consults the country councils via the Landstingens Läkemedelsförmånsgrupps (Pharmaceutical Benefits Group for County Councils) before the final decision is made.

Reimbursement decision drivers include the drug's cost-effectiveness, marginal benefit over alternative treatments, the severity of the disease, and the unmet needs for a new drug. The TLV assesses quality of life (QoL), life expectancy, cost savings, and quality adjusted life years (QALYs). Moreover, cost-effectiveness is analyzed from a societal perspective, which means that all relevant costs and effects of treatment and ill health are considered, regardless of who pays or benefits. For medicines, this means that the analysis of direct costs also takes productivity costs into account for sick leave or increased productivity if the patient can start work again.[9]

In addition to the factors assessed above, the TLV's decisions issue from the Pharmaceutical Benefits Act (2002) as well as the overarching goal of promoting good health and equal access to health care. They take into account three fundamental principles[10]:

- *The cost-effectiveness principle*: The cost of using a medicinal product should be reasonable from a medical, humanitarian, and socioeconomic perspective.
- *The need and solidarity principle*: Those with the most pressing medical needs should have more of

the health care system's resources than other patient groups.
- *The human value principle*: The health care system should respect the equal value of all human life.

All three must be considered and weighed together by the TLV when making its decision on reimbursement.

The TLV's Pharmaceutical Benefits Board votes by simple majority. The final decision may be one of the following: reimbursed, reimbursed with restrictions, reimbursed with conditions, and not reimbursed.[11] Manufacturers are given the opportunity to appeal in the case of the latter three decisions.

The Pharmaceutical Benefits Scheme is a national system and covers the whole country. Reimbursement decisions by the TLV at a national level are adopted by the county councils but with varied flexibility. If the TLV rejects a drug at the national level, a county council may still decide to fund it, based on their needs. County-level drugs committees may also demand additional conditions or restrictions to the reimbursement terms laid down by the TLV or may not recommend the use of the drug.

16.2.3   PRICING PROCESS

As discussed above, the TLV decision is a joint reimbursement and price decision.[12] The manufacturer submits a proposed price as part of its P&R application to the TLV. The prices are assessed based on clinical and cost-effectiveness evidences, through the value-based pricing (VBP) system.

Three core principles underlie the Swedish VBP system[13]:

- A societal perspective that explicitly takes into account economic effects beyond the health sector is adopted. This may include the value of caregivers' time and foregone income, a societal cost often poorly recognized in HTA systems.
- Sweden does not have a national "threshold" QALY value and instead focuses on individuals' willingness to pay for a QALY. This threshold can vary with disease severity.
- Sweden recognizes explicitly that some indications for the same medicine may provide greater health gains than others. This can be a factor in determining whether and at what level to reimburse a new treatment.

The TVL does not negotiate prices. If the price is too high, the board rejects the drug's application. The company may then decide whether it will re-apply, and, in that case, it suggests a lower price.

16.2.4   PHARMACO-ECONOMIC ASSESSMENT

The TLV assesses the cost-effectiveness of a drug in conjunction with its marginal benefit.[14] Thus, the higher the marginal benefit, the higher the price that the TLV considers cost-effective. The requirements of pharmaco-economic evaluations according to TLV guidelines are as follows:

- It should be performed from a societal perspective and should use Swedish data where possible.
- The costs and health effects of using the drug in question should be compared with the most appropriate alternative treatment in Sweden (e.g., the most used).
- The analysis should include the whole patient population to which the subsidy application refers. Separate calculations should be made for different patient groups where the treatment is expected to have different cost-effectiveness.
- Cost-effectiveness analysis is recommended, with QALYs as the measure of effect.
- QALY weightings should be based on methods such as the standard gamble (SG) or time trade-off (TTO) methods.
- All relevant costs associated with treatment and illness should be identified, quantified, and evaluated.
- The time frame for the study shall cover the period in which the main health effects and costs arise.
- Both costs and health effects should be discounted by 3%.
- Methods, assumptions made, and detailed data shall be shown so clearly that the different steps in the analysis are easily followed.
- It is understood that a health economic study that has been peer reviewed and published in an international scientific journal has undergone a form of quality control.

These guidelines are nonbinding and act as a support tool.

16.3 TIME TO MARKET

Per regulation for new drugs, the TLV is obliged to make an overall pricing and reimbursement decision within 180 days.[15] Decision times can be shorter than this, depending on the level of complexity of the submission. For generics, the time is shorter, at 5–30 days.[8]

To reduce the time to market, pharmaceutical companies may submit a P&R application 90 days before a marketing authorization is expected.[16]

16.4 PRICE REGULATIONS

16.4.1 PRICING POLICY FOLLOWING THE MARKETING AUTHORIZATION

Following the marketing authorization, manufacturer sets the price and can apply for inclusion in the Pharmaceutical Benefit Scheme. Non-reimbursed drugs may be priced freely.

16.4.2 EXTERNAL REFERENCE PRICING

External reference pricing was abolished in 2002, and VBP system was formalized in the same year.

16.4.3 INTERNAL REFERENCE PRICING

Internal reference pricing is not applied in Sweden. However, generic substitution and price ceiling for substitutable

pharmaceuticals are implemented. All products containing the same active ingredient are clustered into interchangeable drug groups. In each group, the ceiling pharmacy purchase price for the group is calculated and set by the TLV. Any price lower or the same as this ceiling price within the group is accepted, without further investigations.

To boost competition between the pharmaceutical companies, the TLV implements a process for price decisions concerning substitutable drugs. The manufacturer can apply for an increase or a decrease in drug price, once within the drug groups. If the new price is lower than or similar to the ceiling price within a group of substitutable medicines, the TLV allows both price cuts and price rises, without further investigation. These decisions are made once a month; however, it should be noted that the companies do not know which price their competitors have applied for. As a result, the company that can offer the lowest price will get the majority of the sales because automatic substitution by pharmacies to the cheapest generic is mandatory in Sweden. This creates robust price competition.[17]

16.4.4 PRICE CONTROL AT EX-FACTORY PRICE LEVEL

Price control at ex-factory level is generally not applied in Sweden.[18] However, price cuts have been implemented from 2014 (see Section 16.7.6).

Price reductions are also possible following a reimbursement assessment by the TLV, which can occur in various situations such as the approval of an additional indication and a change in the treatment landscape such as new comparators launched, among others.[19]

16.4.5 PRICE CONTROL AT WHOLESALE LEVEL

Price control at wholesale level is applied for reimbursed medicine by regulating that the wholesalers sell the medicine to the pharmacies at the pharmacy purchase price granted by the TLV.[8] However, in order to gain profit, wholesalers are free to negotiate discounts with manufacturers. These are not regulated by the TLV, and agreements between wholesalers and manufacturers are not public.

16.4.6 PRICE CONTROL AT PHARMACY RETAIL LEVEL

The TLV imposes price control of reimbursed medicine by setting pharmacy mark-ups. Mark-ups are computed based on the pharmacy purchase price and are regressive.[8]

16.4.7 MANDATORY PRICE REDUCTION ON BRAND PRICE AFTER GENERIC/BIOSIMILAR ENTRY

Mandatory price reduction on brand price on generic/biosimilar entry is not applicable in the Swedish system. However, in line with the regulation on generic substitution and the setting of price ceilings for drugs in interchangeable groups, the prices of reimbursed off-patents in the group are cut by 65%, once the pharmacy purchase price of the lowest-priced generic is at least 70% below the price of the off-patent originator.

16.5   REIMBURSEMENT SPECIFICITIES

The Swedish reimbursement system is designed to protect patients from high costs.[20]

The maximum cost for a patient for prescription medicines in the high-cost threshold system is 2,200 SEK during a 12-month period. Hence, during a 12-month period, a patient can pay maximum 2,200 SEK as copayment. The level of copayment decreases with increasing overall spending. For costs up to 1,100 SEK, the patient pays 100% of the cost. In the other extreme, if the cost is more than 5,400 SEK, it is 100% reimbursed and the patient does not pay any copayment.

If a patient has bought medicines on prescription for 2,200 SEK within a 12-month period, then he or she does not pay more for his or her medicines during the remaining time in that period.

The TLV has the possibility to decide that a pharmaceutical will receive a high-cost protection for only a particular field of application or a particular patient group, a so-called restricted subsidy.

Patients can opt to pay for drugs that are not reimbursed by the Swedish National Health Service.

16.6   CHARACTERISTICS OF PUBLIC TENDERING

Public procurement of medicines used in hospitals is carried out by the county councils. Pharmaceutical boards of each county council evaluate and accept tenders based on expert groups' recommendation. Counties can get together in order to negotiate a better discount (due to the larger volume). Public tendering is not applied to ambulatory care.[21]

16.7   EXPENDITURE CONTROLS

16.7.1   DISCOUNTS/REBATES

Reimbursed medicines are priced according to the Pharmaceutical Benefits Act (2002), and no further negotiations or discount of the price take place for outpatient drugs.[22] However, it is common that county councils are given discounts on medicines used in hospitals through negotiations with manufacturers.

16.7.2   CLAWBACK

Clawbacks are not used in Sweden.[8]

16.7.3   PAYBACK

Paybacks are not used in Sweden.[8]

16.7.4   PRICE-VOLUME AGREEMENTS

Price-volume agreements can be used in Sweden in the hospital setting.[8] The county councils are free to negotiate agreements for hospital drugs, especially high-cost medicines, with manufacturers.

16.7.5   OTHER MARKET ACCESS AGREEMENTS

Coverage with evidence development is used in Sweden. Manufacturers are required to submit data on use or cost-effectiveness, depending on the type of uncertainty.[23]

16.7.6   PRICE FREEZES AND CUTS

Price cuts have been implemented in Sweden. From 2014, off-patent originals and generics that had been marketed for 15 years and not covered by any price ceiling will be subject to a 7.5% price cut. Bi-annual price cuts will continue moving forward.

Price freezes are not applicable in the country.[8]

16.8   POLICIES TARGETED AT WHOLESALERS, PHARMACISTS, PHYSICIANS, AND PATIENTS

16.8.1   WHOLESALER MARK-UP

Wholesaler marks-ups are unregulated in Sweden and fixed after negotiations with manufacturers. The average margin for wholesalers is 3.5% of the pharmacy purchase price.

16.8.2   PHARMACY MARK-UP

In Sweden, pharmacy mark-ups are regulated and are regressive. The average margin for pharmacies is 21% of the pharmacy retail price.[8]

16.8.3   GENERIC SUBSTITUTION

Generic substitution is mandatory for pharmaceuticals that contain the same substance, the same formulation, and are deemed comparable by the MPA.[8] The pharmacy has to dispense the least expensive generic. Prices can be changed every month, and competition is high. However, the doctor or the patient can decide against the substitution. In the latter case, the patient has to pay the extra price difference.

16.8.4   INTERNATIONAL NONPROPRIETARY NAME (INN) PRESCRIBING

Prescribing with INN is not mandatory in Sweden.[8]

16.8.5   PRESCRIPTION GUIDELINES

Sweden has treatment guidelines on the national and regional levels. Prescription guidelines are indicative.[8] There are no sanctions against doctors for not following the guidelines, as long as they are not malpracticing.

16.8.6   MONITORING OF PRESCRIBING BEHAVIOR

Prescription behavior is monitored at the council level via the "workplace code." Physicians are obliged to indicate their workplace code on each prescription for it to be reimbursed. Prescription patterns can be monitored using this code.[24]

16.8.7 PHARMACEUTICAL BUDGETS DEFINED FOR PHYSICIANS

Target budgets are set by councils for physicians.[25] Target budgets are linked to compliance, with prescribing recommendations drawn at the regional level.

16.8.8 PRESCRIPTION QUOTAS

Prescription quotas are not used in Sweden.[8]

16.8.9 FINANCIAL INCENTIVES FOR PHYSICIANS

Physicians may be financially incentivized to follow prescription guidelines and budgets.[26] Incentives vary from one council to another. Some county councils have incentive agreements where both adherence to the budget and prescription targets result in a reward.

16.8.10 FINANCIAL INCENTIVES FOR PHARMACISTS

Financial incentives are not used in Sweden for pharmacists.

16.8.11 COPAYMENT FOR PATIENTS

Patient copayments depend on their total health care spending. During a 12-month period, the initial spending is paid 100% by the patient, subsidy level then increases by increment, and costs more than 5400 SEK are fully subsidized. On average, patients pay 29% of the costs for outpatient medicines.

REFERENCES

1. TLV. Accessed July 2016, from: http://www.tlv.se/in-english-old/organisation/our-mission/
2. The Swedish Council on Technology Assessment in Healthcare. Accessed July 2016, from: http://www.sbu.se/en/About-SBU/
3. The Medical Products Agency. Accessed July 2016, from: http://www.lakemedelsverket.se/english/
4. Socialstyrelsen. Accessed July 2016, from: http://www.social-styrelsen.se/english
5. The Ministry of Health and Social Affairs. Accessed July 2016, from: http://www.government.se/sb/d/573
6. Health Systems in Transition: Sweden. 2012. Accessed July 2016, from: http://hspm.org/countries/sweden25022013/countrypage.aspx#
7. Ispor. Global Healthcare Systems Road Map—Sweden Pharmaceutical. Last updated: May 2009. Accessed from: http://www.ispor.org/htaroadmaps/Sweden.asp
8. TLV. Guide for companies when applying for subsidies and pricing for pharmaceutical products. 2012. Accessed July 2016, from: http://www.tlv.se/Upload/English/ENG-guide-for-companies.pdf
9. TLV. Health Economics. Accessed October 2015, from: http://www.tlv.se/In-English/medicines-new/health-economics/
10. TLV. The reimbursement decision process. Accessed October 2015, from: http://www.tlv.se/In-English/medicines-new/apply-for-a-price-or-reimbursement/the-decision-process/
11. TLV. Types of reimbursement. Accessed October 2015, from: http://www.tlv.se/In-English/medicines-new/pricing-and-reimbursement-of-medicines/types-of-reimbursement/
12. TLV. Guide for companies when applying for subsidies and pricing for pharmaceutical products. 2012. Accessed October 2015, from: http://www.tlv.se/Upload/English/ENG-guide-for-companies.pdf
13. Persson U. Value Based Pricing in Sweden: Lessons for Design? Accessed July 2016, from: https://www.ohe.org/publications/value-based-pricing-sweden-lessons-design
14. TLV. General guidelines for economic evaluations from the Pharmaceutical Benefits Board (LFNAR 2003:2). Accessed October 2015, from: http://www.tlv.se/Upload/English/Guidelines-for-economic-evaluations-LFNAR-2003-2.pdf
15. TLV. Processing times. Accessed October 2015, from: http://www.tlv.se/In-English/medicines-new/apply-for-a-price-or-reimbursement/processing/
16. TLV. Guide for companies when applying for subsidies and pricing for pharmaceutical products. 2012. Accessed July 2016, from: http://www.tlv.se/Upload/English/ENG-guide-for-companies.pdf
17. TLV. The Swedish Pharmaceutical Reimbursement System. 2007. Accessed October 2015, from: http://www.tlv.se/Upload/English/ENG-swe-pharma-reimbursement-system.pdf
18. Carone G, Schwierz C, Xavier A. Cost-containment policies in public pharmaceutical spending in the EU. European Economy—Economic Papers 461. September 2012. Accessed July 2016, from: http://ec.europa.eu/economy_finance/publications/economic_paper/2012/pdf/ecp_461_en.pdf
19. TLV. Pharmaceutical reviews. Accessed October 2015, from: http://www.tlv.se/In-English/medicines-new/the-pharmaceutical-review/
20. TLV. High-cost threshold. Accessed October 2015, from: http://www.tlv.se/In-English/medicines-new/the-swedish-high-cost-threshold/how-it-works/
21. PHIS. PHIS Hospital Pharma Report: Sweden. 2009. Accessed October 2015, from: http://whocc.goeg.at/Literaturliste/Dokumente/CountryInformationReports/V5%20PHIS%20Hospital%20Pharma%20report%20SE_final%20version.pdf
22. Vogler S, Zimmermann N, Habl C, Piessnegger J, Bucsics A. Discounts and rebates granted to public payers for medicines in European countries. *Southern Med Review*, 5(1):38–46. 2012.
23. Ferrario A, Kanavos P. Managed entry agreements for pharmaceuticals: The European experience. EMiNet, Brussels, Belgium. April 2013.
24. PPRI. Sweden. Accessed October 2015, from: http://whocc.goeg.at/Literaturliste/Dokumente/CountryInformationReports/Sweden_PPRI_2007.pdf
25. PPRI. Sweden. Accessed October 2015, from: http://whocc.goeg.at/Literaturliste/Dokumente/CountryInformationReports/Sweden_PPRI_2007.pdf
26. PPRI. Sweden. Accessed October 2015, from: http://whocc.goeg.at/Literaturliste/Dokumente/CountryInformationReports/Sweden_PPRI_2007.pdf

17 United Kingdom

17.1 STAKEHOLDERS

17.1.1 National Pricing and Reimbursement Decision Makers

In the United Kingdom, the Department of Health (DH) is the national pricing and reimbursement decision maker.[1]

17.1.2 National Health Technology Assessment (HTA) Agencies

The United Kingdom has three regional bodies responsible for providing guidance on the cost-effective use of medicines:

- *England and Wales*: The National Institute for Health and Care Excellence (NICE), which conducts appraisals and develops guidelines.
- *Scotland*: The Scottish Medicines Consortium (SMC), which assesses all new medicines at launch as to whether they are cost-effective for use in Scotland.
- *Wales*: All Wales Medicines Strategy Group (AWMSG) provides guidance to the Minister of Health and Social Services in Wales on strategic medicines management and prescribing. The National Coordinating Centre for Health Technology Assessment manages and develops the National Health Service (NHS) HTA program.[2]

17.2 PRICING AND REIMBURSEMENT POLICIES

17.2.1 Overview of the System

The NHS is responsible for providing universal health care and is funded by the central UK government, primarily via general taxation and by national insurance contributions from residents of the United Kingdom.

Once a pharmaceutical company has obtained marketing authorization and pricing approval, it is free to launch the medicine in the United Kingdom and in most cases will be granted automatic full reimbursement on the NHS. Hospitals may be able to purchase medicines under contract at a discount to the NHS list price. However, generics, in vitro diagnostics, unlicensed products sold on an individual patient supply basis, dental anesthetics, and over-the-counter (OTC) drugs are not reimbursed.

The Pharmaceutical Price Regulation Scheme (PPRS) governs the prices of branded prescription drugs sold to the NHS, including branded generics, vaccines, in vivo diagnostics, blood products, dialysis fluids, biotech products, and branded drugs supplied through a tendering process or via a central or local contract. The PPRS results from a voluntary non-contractual agreement negotiated between the DH and the Association of the British Pharmaceutical Industry (ABPI).[3]

In the United Kingdom, applications for pricing and reimbursement are not separated, and appraisals of pharmaceuticals can be via one of these two processes:

- Multiple-Technology Assessment (MTA): These assessments typically take around 1 year to complete and involve looking at more than one technology or multiple indications for the same technology.
- Single-Technology Assessment (STA): NICE evaluates single technologies for single indications, with guidance available within six months of a technology being granted marketing authorization.

NICE compares health care interventions by using the incremental cost-effectiveness ratio (ICER), which quantifies the cost per unit of benefit gained (e.g., quality adjusted life years [QALYs]) from using one treatment versus another. According to the value of ICER obtained, medicines can be grouped into three categories based on the possible types of recommendations[3]:

- NICE recommendation is likely to be positive if ICER <£20,000
- NICE recommendation is not predictable if ICER is £20,000-£30,000
- NICE recommendation is likely to be negative if ICER >£30,000

17.2.2 Reimbursement Process

In England, reimbursement is theoretically granted to all products once they have received marketing authorization, except for drugs included in the negative list or the gray list,[3] in Scotland, where health boards decide whether or not to fund a drug based on the recommendations of the SMC, which takes precedence over NICE guidance.[3] However, in Wales, Welsh Local Health Boards (LHBs) rely on NICE guidance when deciding on funding for a particular drug. In addition, for products that have not been evaluated by NICE, LHBs must take into account any guidance produced by the AWMSG.[3]

In Northern Ireland, NICE guidance is taken into account by Health and Social Care Trusts when making funding decisions.[3]

Reimbursement prices for generic medicines (Category A or Category M, with category M holding the majority of generics) are set and published monthly in the Drug Tariff (DT) and are reimbursed at the prices listed in Part VIIIA of the DT.[3]

Reimbursement prices of category A medicines are adjusted on a monthly basis, with pharmacists negotiating discounts on the tariff price in order to earn a profit. The DH adjusts tariff prices quarterly, based on pricing and cost data submitted by manufacturers. Pharmacists earn profit by negotiating discounts on the tariff price.[3]

17.2.3 Pricing Process

The PPRS controls the prices of branded prescription medicines supplied to the NHS by regulating the profits that companies can make on their sales to the NHS. Within this framework, there are elements of free pricing, price negotiations and procurement, and statutory pricing for companies that do not join the voluntary PPRS. A company supplying the NHS with branded medicines, which has not joined the PPRS, falls under the Health Service Branded Medicines.

The PPRS does not set prices at the manufacturer level. Rather, it sets the NHS list price, that is, the reimbursement price, which includes wholesaler and pharmacy margins, while the prices of generic medicines purchased by the pharmacy are set by the market.

All manufacturers seeking to launch a new drug are required to submit the following information to the DH, giving at least 28 days' notice of their intention[3]:

- The proposed launch price (NHS list price)
- Summary of product characteristics (or draft thereof)
- The anticipated level of uptake and proposed NHS list price in each of the first 5 years

The members of PPRS with total NHS sales (at list price) in excess of £50 million per annum must provide an Annual Financial Return (AFR) to the DH, which is used to calculate a manufacturer's permitted level of profitability.

The AFR must contain the following information[3]:

- Gross sales (at NHS list price) for each product for each quarter
- Net sales (gross sales less any applicable rebates, discounts, and so on) per product per quarter
- The information should be further broken down by distribution channel
- A breakdown of the manufacturer's costs (e.g., R&D costs and marketing costs) and amount of capital employed in the United Kingdom

Manufacturers with sales below £50 million per annum are not required to provide the DH with any additional financial information.[3]

Value-based pricing (VBP) had been stated to be implemented in the PPRS from January 2014. However, England failed to implement the VPB using the Multicriteria decision analysis (MCDA) method.

Further, under the new PPRS, new drugs can be "priced at the discretion" of the manufacturer in order to achieve value for money for the NHS. This is achieved as a part of the NICE technology appraisal under a provision called Patient Access Schemes (PAS).[3]

17.3 TIME TO MARKET

For new medicines, the average time between the dates of the European Medicines Agency (EMA) marketing authorization and the "accessibility" date is 118 days.[4] For generics, the market access is immediate after obtaining marketing authorization.[3]

17.4 PRICE REGULATIONS

17.4.1 Pricing Policy Following the Marketing Authorization

After obtaining the marketing authorization from the EMA or from the Medicines and Health care Products Regulatory Agency (MHRA) at the national level, manufacturers are free to set prices. However, prices of branded prescription medicines supplied to the NHS are controlled through the PPRS by regulating the profits that companies can make on their sales to the NHS.[3] These controls apply only to those manufacturers whose annual NHS sales exceed £50 million via the following mechanisms:

- The allowable return on capital (ROC) is fixed at 21% of the annual capital employed by the company in the United Kingdom
- A margin of tolerance (MOT) is associated with the allowable ROC:
  - Manufacturers are permitted to retain profits of up to 150% of the allowable ROC
  - The MOT is not available to a manufacturer, in any year in which it has implemented an authorized price increase

17.4.2 Reference Pricing

External and external reference pricings are not applied in the United Kingdom.[5,6]

17.4.3 Price Control

Price control at ex-factory price[7] and at wholesale[7] levels is applied for reimbursed medicines. At pharmacy retail level,[7] price control is applied for reimbursed medicines and retailers are able to set their own prices competitively and can choose to sell at a price above or below the retail price recommended by the manufacturer. The recommended retail price includes VAT and a margin for the pharmacist.[3]

17.4.4 Mandatory Price Reduction on Brand Price after Generic/Biosimilar Entry

There is no mandatory price reduction on brand price after generic or biosimilar entry in the United Kingdom market.

17.5   REIMBURSEMENT SPECIFICS

Products in the negative list are excluded from reimbursement. Thus, while general practitioners (GPs) are permitted to write a private prescription for such products, they must not charge an NHS patient for the consultation, but the patient is obliged to pay the full cost of the private prescription out of pocket.

The negative list contains 17 therapeutic categories that do not have a therapeutic or a clinical advantage over other, cheaper drugs, non-reimbursed OTCs, and products for which the cost to the NHS is not justifiable or that are not considered as a priority.

The "gray list" called the Selected List Scheme (SLS) restricts access to certain medicines. GPs are allowed to prescribe a product in the gray list for only certain indications and for only some groups of patients. The positive list is not applied in the UK system.[6]

17.6   CHARACTERISTICS OF PUBLIC TENDERING

There has been a rise in the use of public tendering, a process that is geared toward increasing price competition and achieving reductions in purchase prices of pharmaceuticals.

17.6.1   Applied to Hospital Care

In terms of hospital care, the types of medicines purchased through tendering procedures are vaccines, pharmaceuticals used in pandemic plans, as well as branded pharmaceuticals and generics prescribed against non-communicable diseases. Price is the most important criterion for winning a tender, among other criteria, including the availability of the medicine. Public tendering in the United Kingdom is the most economically advantageous tender.[6]

17.6.2   Applied to Ambulatory Care

Public tendering is mostly used in hospital settings, although an increasing tendency to use it in ambulatory care is possible in the foreseeable future.[6]

17.7   EXPENDITURE CONTROLS (SUPPLY SIDE)

17.7.1   Discounts/Rebates

The NHS list price of branded medicines includes a margin for distribution through wholesalers.

For branded medicines and generics, the level of discount is negotiated between the manufacturer and wholesaler and pharmacy and varies over time, from product to product, and from company to company. There are no restrictions on the type of discount (within the constraints of competition law).[6]

17.7.2   Clawback

In England, for both branded and generic drugs, the DH recoups from pharmacists a portion of the discounts that they have negotiated with suppliers, under the terms of the Community Pharmacy Contract. This clawback is calculated according to a sliding scale published in Part V of the DT and varies according to a pharmacy's total monthly reimbursement. It is based on information obtained from pharmacy invoices. Drugs that are excluded from the clawback pharmacist arrangements are included in the zero discount list of Part II of the DT. These drugs are represented by controlled drugs, certain hazardous chemicals, cytotoxic or cytostatic items, and drugs requiring cold-chain storage.[3]

In Scotland, the NHS is able to clawback money from pharmacists on branded prescription drug discounts negotiated with suppliers. The clawback is calculated according to a sliding scale, to enable the overall clawback target to be increased to 6.24%. Drugs included on the zero discount list are exempted from the clawback arrangements, and the pharmacy clawback on generic drugs has been abolished.[3]

17.7.3   Payback

In the United Kingdom, there are no payback arrangements in place for the industry.[3]

17.7.4   Price-Volume Agreements

In the United Kingdom, the price-volume agreement is one of the various "risk sharing" schemes used. Volume agreement schemes trigger refunds from the manufacturers if pre-agreed sales or volumes are exceeded. Refunds may be in form of a lowering of reimbursed prices. Payback policies are another form of price-volume agreements, which are mostly on the level of total pharmaceutical spending.[3]

17.7.5   Other Market Access Agreements

There are no specific funding or reimbursement schemes for high-cost or innovative medicines, but two special arrangements exist:

- Patient Access Schemes (PAS): Proposals from pharmaceutical companies offering discounts or other arrangements to reduce the cost of a drug to the NHS without changing the formal NHS list price. PAS proposals can only be made in the context of a NICE appraisal.
- Cancer Drugs Fund: £200 million (€240 million) for 3 years from April 2011 to help patients get the medicines that their doctors recommend.

In order to secure NHS reimbursement of their product, manufacturers can propose a Patient Access Scheme (PAS), of which different types exist as follows[3]:

- Rebate for non-responders
- Payment cap
- Discount
- First treatment cycle free

- Initial treatment cycle free
- Fixed price
- Evidence-based discount

17.7.6 PRICE FREEZES AND CUTS

The 2009 PPRS scheme introduced stepped price cuts on branded pharmaceuticals. It outlined a 3.9% price cut, which was implemented in February 2009, followed by another 1.9% cut in January 2010. However, the price cut imposed on PPRS drugs in 2009 did not cover products authorized for sale after the beginning of 2009.[6]

17.8 POLICIES TARGETED AT WHOLESALERS, PHARMACISTS, PHYSICIANS, AND PATIENTS

17.8.1 WHOLESALER MARK-UP

The NHS list price of medicines includes a margin for distribution through wholesalers, where the level of discount is negotiated between the manufacturer and wholesaler on the one hand, and between the wholesaler and pharmacy on the other hand.

This varies over time, from product to product and from company to company. In 2011, the average wholesaler margin was equal to 12.5% PPP (pharmacy purchase price).[6]

17.8.2 PHARMACY MARK-UP

In England, a framework is negotiated with the Pharmaceutical Services Negotiating Committee, the organization representing the interests of community pharmacies,[6] in which pharmacies' margins are determined by the difference between NHS reimbursement price and the actual pharmacy purchase price, and pharmacists receive fees and allowances for their services.[8]

In Scotland, pharmacy margins on branded and generic drugs are not regulated. Instead, they are subject to negotiation between pharmacists and suppliers.[6]

17.8.3 GENERIC SUBSTITUTION

A pharmacist may choose to substitute a cheaper (generic) medicine with the same active ingredient(s) for another, usually a brand medicine, in what is referred to as generic substitution. However, this is disallowed in the United Kingdom, where pharmacists are expected to dispense what is written in the prescription.[6]

Substitution may occur in hospital settings but only according to well-defined, locally developed formulary arrangements.[6]

17.8.4 INTERNATIONAL NONPROPRIETARY NAME (INN) PRESCRIBING

The INN is indicative in the United Kingdom and is a key tool to improve prescribing behavior.[6]

17.8.5 PRESCRIPTION GUIDELINES

Pharmaceutical companies are bound by a voluntary code of practice on prescription promotion set by the Association of the British Pharmaceutical Industry (ABPI). This code is accepted by most UK pharmaceutical companies, including non-ABPI members. Pharmaceutical companies that are found to be in breach of the code have to undertake to cease the promotional activity in question and take steps to avoid a similar breach in the future and pay an administrative fine. Serious breaches can lead to reprimand and expulsion from the ABPI.[6]

17.8.6 MONITORING OF PRESCRIBING BEHAVIOR

Monitoring of doctors' prescribing is operational at the PCT level.[10]

17.8.7 PHARMACEUTICAL BUDGETS DEFINED FOR PHYSICIANS

Pharmaceutical budgets are a cost-containment measure of third-party payers, where the maximum amount of money to be spent on pharmaceuticals in a specific region and period of time is fixed exante.[9]

Budgets for primary care under the NHS are increasingly being devolved to groups of general practices (Primary Care Trusts [PCTs]) and to individual general practices themselves.

Under the policy of practice-based commissioning, general practitioners (GPs) have been able to assume control of their own budgets from their local PCT.[6]

17.8.8 FINANCIAL INCENTIVES FOR PHYSICIANS

A financial incentive is currently in place for physicians in the United Kingdom, which appears to have contributed to an increase in the generics market share.[9]

17.8.9 COPAYMENT FOR PATIENTS

Health care coverage in the United Kingdom is universal and free. Both inpatient and outpatient coverages are free. However, patients in England pay a fixed charge for outpatient prescription drugs. Patients who are not eligible for an exemption must pay a standard charge of £8.40 per item dispensed. Patients who require a high number of prescriptions can make savings by purchasing a Prescription Pre-Payment Certificate (PPC), which enables them to access unlimited prescriptions for up to 1 year for a one-off fixed fee:

- 3-month PPC: £29.10
- 12-month PPC: £104.00

In Scotland, prescription charges were abolished from April 1, 2011, but before this date, patients had to pay £3.00 per item dispensed.[6]

The Welsh Assembly abolished prescription charges in 2007. However, to qualify, prescriptions must be written in English and Welsh script in order to discourage "prescription tourism" into regions that border England.[6]

In Northern Ireland, prescription charges were abolished from April 1, 2010. However, in September 2012, the Northern Irish Health Minister proposed a reintroduction of the charge so as to use it as new fund for high-cost drugs for the treatment of conditions such as cancer, arthritis, and psoriasis.[6]

REFERENCES

1. Pharmaceutical Pricing and Reimbursement Information. Report. 2008. Gesundheit Österreich GmbH / Geschäftsbereich ÖBIG, Vienna, Austria. Available from: https://ppri.goeg.at/Downloads/Publications/PPRI_Report_final.pdf (accessed July 10, 2016).
2. ISPOR Roadmap, United Kingdom, 2008. Available from: http://www.ispor.org/htaroadmaps/uk.asp (accessed July 10, 2016).
3. The Pharmaceutical Price Regulation Scheme 2014. Department of Health and the Association of the British Pharmaceutical Industry. December 2013. London, UK. Available from: https://www.gov.uk/government/uploads/system/uploads/attachment_data/file/282523/Pharmaceutical_Price_Regulation.pdf (accessed July 10, 2016).
4. Efpia. Patient WAIT report, 2011.
5. Leopold C, Vogler S, Mantel-Teeuwisse A, de Joncheere K, Leufkens H.G.M, Laing R. Differences in external price referencing in Europe. A descriptive overview. *Health Pol.* 2012; 104(1):50–60.
6. Carone G, Schwierz C, Xavier A. Cost-containment policies in public pharmaceutical spending in the EU.(European Economy. Economic Papers. 461. Sep 2012. Brussels. PDF. 62pp). Available from: http://ec.europa.eu/economy_finance/publications/economic_paper/2012/pdf/ecp_461_en.pdf (accessed July 10, 2016).
7. Vogler, S. The impact of pharmaceutical pricing and reimbursement policies on generics uptake: Implementation of policy options on generics in 29 European countries-an overview. In: *Generics and Biosimilars Initiative Journal* (GaBI Journal), 1(2):93–100. Available from: http://gabi-journal.net/the-impact-of-pharmaceutical-pricing-and-reimbursement-policies-on-generics-uptake-implementation-of-policy-options-on-generics-in-29-european-countries%E2%94%80an-overview.html (accessed July 10, 2016).
8. Organisation for Economic Co-operation and Development (OECD). *Pharmaceutical Pricing Policies in a Global Market.* Paris, France: OECD Publishing, 2008. 46–49.
9. Espín J, Rovira J. Analysis of differences and commonalities in pricing and reimbursement systems in Europe. June; Andalusian School of Public Health; Commissioned by the European Commission, Directorate-General Enterprise. Available from: http://whocc.goeg.at/Literaturliste/Dokumente/FurtherReading/EASP%20Report%202007_Analysis%20of%20differences%20and%20commonalities.pdf (accessed July 10, 2016).
10. Scoggins A, Tiessen J, Ling T, Rabinovich L. 2006. Prescribing in primary care. Technical report, RAND Corporation, Santa Monica, CA. Available from:https://www.nao.org.uk/wp-content/uploads/2007/05/TR443_3C.pdf (accessed August 10, 2016).

18 Belgium

18.1 STAKEHOLDERS

18.1.1 NAMES OF NATIONAL PRICING AND REIMBURSEMENT DECISION MAKERS

The National Institute for Health and Disability Insurance (*Institut Nationald'Assurance Maladie-Invalidité* [INAMI], *Rijksinstituut voor ziekte-en invaliditeitsverzekering* [RIZIV]) is responsible for the organization and the reimbursement of health care expenses.[1] This federal institution establishes the rules for the reimbursement and determines the tariffs of the health care services (the so-called nomenclature) and pharmaceuticals. It organizes, manages, and supervises the implementation of the compulsory health insurance and inspects both the sickness funds and the health care providers to see whether they correctly apply the rules of the health care and health insurance system.

The Minister of Economic Affairs is responsible of fixing the maximum ex-factory price of drugs and similar products.[1] The Minister is advised by the Committee of Pricing for Pharmaceutical Specialties (*Commission des Prix des Spécialités Pharmaceutiques* [CPSP], *Prijzencommissie voor de Farmaceutische Specialiteiten* [PFS]). The *Commission de Remboursement des Médicaments* (CRM) (*Commissie voor Tegemoetkoming Geneesmiddelen* [CTG]) is a committee of experts and stakeholders in pharmaceutical reimbursement set up within the INAMI/RIZIV.[1] CRM/CTG is responsible for appraising reimbursement proposals and advising the Minister of Health and Social Affairs about the reimbursement of medicines. The legal missions of the CRM/CTG are as follows:

- Formulating proposals to register pharmaceuticals on the list of reimbursable pharmaceuticals
- Advising, on request of the Minister, on political aspects related to pharmaceuticals reimbursement
- Formulating proposals for the Committee for Health Care Insurance of the INAMI/RIZIV of interpretation rules regarding the pharmaceuticals reimbursement

Before adding a pharmaceutical on the positive reimbursement list, the Minister of Budget is entitled to advise the Minister of Health and Social Affairs on budgetary considerations.

18.1.2 NAMES OF NATIONAL HEALTH TECHNOLOGY ASSESSMENT AGENCIES

The above-mentioned CRM plays an important role in reimbursement decisions. It is supported by internal experts (on the pay-roll of the INAMI) who assess the quality of the reimbursement files submitted by the industry.

Another federal body, the *Kenniscentrum- Centre d'Expertise* (KCE) also conducts health technology assessments (HTAs), but either on own initiative or upon request of the Minister or INAMI. Most HTAs conducted by the KCE are not on medicines but on medical devices and biological tests.

18.1.3 NAMES OF OTHER KEY STAKEHOLDERS (REGIONAL/LOCAL LEVEL)

The Dutch-, French-, and German-speaking Community Ministries of Health are responsible for health promotion; maternity and child health services; different aspects of elderly care; the implementation of hospital accreditation standards; and the financing of hospital investment.[1]

Reimbursement of the health care costs is managed by the sickness funds (the so-called "mutualities"). Citizens are free to choose among six private, nonprofit, and one public sickness funds. Private-for-profit health insurance companies accounts for only a relatively small part of the complementary health insurance market.[1]

18.2 PRICING AND REIMBURSEMENT POLICIES OVERVIEW

18.2.1 OVERVIEW OF THE SYSTEM

Belgium has a compulsory national health insurance that covers almost the whole population.[1] There are a number of health insurance schemes, with the three major schemes covering salaried workers, the self-employed, and civil servants and several smaller schemes covering specific industries.[2] Nearly 75% of the population has a supplementary private insurance that covers procedures and pharmaceuticals not covered by the national insurance.

Responsibility for health care policy is shared between the federal regional governments. At the central level, these are the Federal Public Service (FPS) Public Health; Food Chain Safety and Environment; the FPS Social Affairs; and the INAMI/RIZIV. They are responsible for the regulation and financing of the compulsory health insurance; the determination of hospital accreditation criteria; the financing of hospitals and heavy medical care units; the legislation covering different professional qualifications; and the registration of pharmaceuticals and their price control.[1] The regions are responsible for the organization and administration of primary care (including general practitioner [GP] services), and have a central role in the development of hospital care, long-term care, and disease prevention policies.[2] GPs can be considered gatekeepers to the secondary care.

The Federal Agency for Medicines and Health Products (*Agence Fédérale des Médicaments et des Produits de Santé* [AFMPS], *Federaal Agentschap voor Geneesmiddelen en Gezondheidsproducten* [FAGG]) is responsible for the quality, safety, and efficacy of drugs on the Belgian market. Market authorization is also granted via the centralized procedure by the European Medicines Agency (EMA).[1]

18.2.2 REIMBURSEMENT PROCESS

Only inpatient and outpatient drugs included on the positive list of reimbursement (i.e., appendix of Royal Decree of 21 December 2001) are covered by the compulsory health insurance.[1] In order to place a drug on the positive reimbursement list, a pharmaceutical company can submit a drug reimbursement request for a licensed medicine to the CRM/CTG.

The reimbursement pathway depends on the "Class" claim. A Class 1 procedure is restricted to drugs with claimed added therapeutic value, Class 2 is for drugs with similar or analogous therapeutic value ("me too drugs"), and Class 3 includes generics. For Class 1, the experts and the CRM/CTG first evaluate the company's claim as compared with the standard alternative therapy with regard to[1] the following:

- Efficacy
- Effectiveness
- Safety
- Comfort (the extent to which the use of the drug by the provider and/or the patient improves administration comfort and/or prevents errors related to drug use)
- Applicability (the extent to which the drug characteristics, for example, contraindications, limit the drug use for certain groups of patients and/or require special precautions)

Only Class 1 submissions involve also pharmacoeconomic assessment from the perspective of the INAMI/RIZIV. Receiving a Class 1 allows the negotiation for a price premium. EMA-designated orphan drugs represent a separate class, which is assessed as Class 1, but without the mandatory pharmacoeconomic assessment. Biosimilar drugs follow a pathway similar to Class 2. For both Class 2 and biosimilar drugs reimbursement basis cannot exceed those of the reference drugs.[1]

In order to appraise the reimbursement request file, the CRM/CTG is entitled to appeal to external and/or internal experts.[1] The drug reimbursement proposals are assessed by the experts and appraised by the CRM/CTG members and subsequently voted during CRM/CTG meetings. The positive or negative reimbursement proposal is transferred to the Ministry of Health and Social Affairs within a limit of 150 days. The Minister is responsible for the final decision, which is to be taken before 180 days. The Minister is entitled to deviate from the CRM/CTG proposal for social and/or budgetary reasons.

The Minister of Social Affairs makes the positive reimbursement decision by means of a Ministerial Decree in the Official Journal (Moniteur belge—Belgische Staatsblad).[1]

In principle, positive and negative decisions taken by the Minister are published on the INAMI/RIZIV website. Information on drug prices and reimbursement are compiled in the Commented Drug Directory (*Répertoire Commenté des Médicaments—Gecommentarieerd Geneesmiddelen Repertorium*) and the Memento-Pharma.[1]

Reimbursement decision is made within 180 days. Otherwise, the applicant reimbursement request is enforced.[1]

18.2.3 PRICING PROCESS

Prices of both reimbursed and nonreimbursed prescription drugs are subject to controls.[2] In case of reimbursed drugs, the Minister for Economic Affairs is responsible for setting the maximum ex-factory prices. Manufacturers submit a pricing application to the Price Department (Service des Prix) of the Federal Public Service for Economic Affairs (Service Public Fédéral Economie) at the same time as a separate reimbursement application to the INAMI.[2]

Manufacturer selling prices (MSP) for Class 2 products are set by the CPSP/PFS in relation to prices abroad and prices of comparator products. Maximum prices for Class 1 products are eligible for a premium and to be set above comparator prices.[1]

The proposed premium MSP must be supported by a breakdown of production (or importation), transportation and research and development (R&D) costs, personnel costs, sales and marketing expenses, and overheads.[2] The manufacturer is permitted to add a maximum markup of 5% (for domestically produced medicines) or 10% (for imported medicines) above these costs.[2]

The Commission makes a recommendation, which is then forwarded to the Minister for Economic Affair and the Minister determines the maximum MSP of the new drug.

For new nonreimbursed drugs, manufacturer has to notify the desired price to the Price Department of Federal Public Service (FPS) Economie and submit the same kind of information as for reimbursed drugs.[2] For new formulations, dosages, and so on of nonreimbursed prescription drugs, the MSP is set by the Minister for Economic Affairs, on the advice of the Permanent Price Commission (*Comité Permanent de la Commission pour la Régulation des Prix*).[2]

In accordance with the EU Transparency Directive, the maximum pricing decision must be communicated to the applicant within 90 days. That price can be further reduced via negotiation between the CRM/CTG and the company. The manufacturer can request a review of the decision while justifying it with extra health-economic analyses.[2] In general, the reimbursement basis is equal to the pharmacy retail price.[1]

18.3 TIME TO MARKET

Time to market for innovative and costly drugs may be extended in Belgium, mostly because pharmaceutical companies may choose to delay the application in this country in anticipation of relatively low pricing for their drug.[3,4]

18.4 PRICE REGULATIONS

18.4.1 PRICING POLICY FOLLOWING THE MARKETING AUTHORIZATION

Prices of both reimbursed and nonreimbursed drugs are subject to controls.[2] Only Class 1 medicines are eligible for price premium. Price is set by the CPSP/PFS. That price can be further reduced via negotiation between the CRM/CTG and the company.[1]

18.4.2 EXTERNAL REFERENCE PRICING

External reference pricing (ERP) is applied for Class 2 drugs as a supportive criterion. Either the average price in the reference countries (26 EU MS) or the price in the country of origin is used.[5] ERP was used as a criterion for price cuts, which were introduced in the 2013 health care budget for reimbursed patented medicines that have been at least 5 years on the market. For these drugs, prices were compared to those in six European countries (Austria, Finland, France, Germany, Ireland, and the Netherlands).

18.4.3 INTERNAL REFERENCE PRICING

In 2001, a reference reimbursement system (*système de remboursement de référence-referentieterugbetalingssysteem*) was introduced. The cluster definition includes all drugs with the same active ingredient independently of the dosage and administration routes and the reference price is defined as a percentage reduction in the reimbursement basis of the original drug.[1]

18.4.4 PRICE CONTROL AT EX-FACTORY PRICE LEVEL

Ex-factory prices are set by the CPSP/PFS.

18.4.5 PRICE CONTROL AT WHOLESALE LEVEL

None.

18.4.6 PRICE CONTROL AT PHARMACY RETAIL LEVEL

There is no price control, but cost-containment. Pharmacists are required to dispense one of the three "cheapest medicines" for all prescriptions written by INN.[2]

18.4.7 MANDATORY PRICE REDUCTION ON BRAND PRICE AFTER GENERIC/BIOSIMILAR ENTRY

There are successive reductions in reference prices and prescription status of medicines after generic entry.[6]

18.5 REIMBURSEMENT SPECIFICS

Individual reimbursement revision can be a part of ministerial decision on reimbursement modalities. It applies to[1] the following:

- Class 1 products and orphan drugs
- Class 2 or 3 products for which the Minister decided to have a revision (mostly budgetary uncertainty)
- Modifications for which the Minister decided to have a revision

As a part of revisions, real life usage data from Belgium is often requested (e.g., observational studies, registries). Review can take place from 18 months to 3 years after first inscription on list or after modification of conditions.

18.5.1 CHARACTERISTICS OF PUBLIC TENDERING

In 2006, a tendering process was introduced for off-patent drugs with the same active ingredient and similar indications. The manufacturer that bids the lowest price receives the best reimbursement, although the other drugs are still reimbursed, but at a lower level.[1] However, this process was used only once.

18.5.2 EXPENDITURE CONTROLS DISCOUNTS/REBATES

Generics are initially priced up to 52.21% below the price of the original branded product with further price cuts along the life cycle.[2]

18.5.3 CLAWBACK

Various pharmacy clawback measures can be periodically applied.[2]

18.5.4 PAYBACK

Paybacks exist since 2006.[1] A share of revenue is paid back by manufacturers, if a prespecified budget ceiling for public pharmaceutical expenditures is exceeded. It is most often based on an annually approved global budgetary target. Orphan drugs, Category Cx drugs, and blood derivates, as well as drugs subject to Market Access Agreements are not subject to payback.

Because of budgetary instable situation, the following additional taxes were installed[5]:

- *Classic tax*: 6.73% on total revenue.
- *Crisis tax*: 1% on total revenue (will be abolished if budgetary situation improves).
- *Subsidiary tax*: Percentage on total revenue fixed if budget overshoot is expected plus maximum of budget overshoot is set at €100 million.
- *Orphan tax*: Percentage per "revenue slice."

18.5.5 PRICE-VOLUME AGREEMENTS

See below-mentioned section.

18.5.6 OTHER MARKET ACCESS AGREEMENTS

Market access agreements are called in Belgium Art. 81 agreements.[7] They can be concluded between INAMI/RIZIV and a pharmaceutical company by negotiations. Agreements try to link the price of a medicinal product to its specific added value, and no longer to the willingness-to-pay. They involve temporary reimbursement based on conditions set out in the contract. The term of contract is 1–3 years.

Negotiation is possible upon request of Commission of Reimbursement of Medicines or when CRM cannot formulate a proposal. The contract includes the following:

- Facial price
- Terms of compensation for budgetary risks
- Terms of scientific reporting and/or evaluation
- Notification of turnover
- Reimbursement conditions

The contracts can include one or a combination of the following models:

- Budget cap
- Price-Volume
- Fixed amount per unit
- Pay for performance scheme
- Reduction price of other drugs (cross deal)
- Risk sharing in case of two kinds of uncertainties:
  - Clinical uncertainty
  - Budgetary uncertainty

Another pathway is called Early Temporary Authorization or Early Temporary Reimbursement (ETA/ETR).[7] Manufacturers can apply for ETA/ETR at the moment of filing a Marketing Authorization dossier with the EMA and obtain temporary reimbursement in the same indication. This is different from Compassionate Use, where the reimbursement is for an indication not included in the EMA dossier.

18.5.7 PRICE FREEZES AND CUTS

Price cuts occurred in 2013 and were based on ERP as described in Section 8.4.2.[4]

18.6 POLICIES TARGETED AT WHOLESALERS, PHARMACISTS, PHYSICIANS, AND PATIENTS

The public price is composed of the ex-factory price plus wholesaler markup, pharmacist markup, and tax. The ex-factory price is about 50% of the public price.[7] VAT is at 6%.[2]

18.6.1 WHOLESALER MARK-UP

Wholesale markups on reimbursed drugs are €0.35 for MSP greater than 2.33, 15% of the MSP for MSP greater than or equal to 2.33 to €15.33, and €2.30 plus (0.9% of the part of the MSP exceeding €15.33) for MSP above €15.33.[2]

18.6.2 PHARMACY MARK-UP

Pharmacy markups for reimbursed medicines are 6.04% of the MSP for MSP less than or equal to €60.00 or €3.624 plus (2% of the part of the MSP exceeding €60.00) for MSP above €60.00.[2] Pharmacists are also paid a dispensing fee of €4.16 per pack plus a fee of €1.28 for drugs prescribed by International nonproprietary name (INN).

18.6.3 GENERIC SUBSTITUTION

Generic substitution by pharmacists is not allowed in Belgium.[1]

18.6.4 INTERNATIONAL NONPROPRIETARY NAME PRESCRIBING

International nonproprietary name (INN) prescription allows pharmacists to deliver first low-priced drug.[1] Antibiotics and antifungals must be prescribed by doctors under INN.[2]

18.6.5 PRESCRIPTION GUIDELINES

Formal prescription guidelines and information campaigns targeted at physicians are formulated by INAMI and the Medicines Reimbursement Commission.[1,2] The Belgian Health Care Knowledge Centre (KCE) publishes advisory recommendations on drugs.[2]

18.6.6 MONITORING OF PRESCRIBING BEHAVIOR

The Pharmanet system monitors doctors' compliance with the low-cost drug targets.[2] Every physician receives individual feedback on their prescribing patterns and those not complying with the guidelines and/or quotas receives additional support for low-cost prescribing.[1]

18.6.7 PHARMACEUTICAL BUDGETS DEFINED FOR PHYSICIANS

Instead of budgets there are prescribing quotas (see Section 8.6.8).

18.6.8 PRESCRIPTION QUOTAS

Since 2006, prescribing quotas for low-cost drugs (i.e., generics or original drugs whose price decreased) exist.[1] GPs are required that 50% of prescriptions are for low-cost drugs.[2] These quotas vary with the specialty (from 9% for the gynaecologists to 27% for GPsand 30% for dentists).[1] Low-cost drugs are featured in the online drug formulary maintained by the Belgian Centre for Pharmacotherapeutic Information (*Centre Belge d'Information Pharmacothérapeutique* [CBIP]).[2]

18.6.9 Financial Incentives for Physicians

None.

18.6.10 Financial Incentives for Pharmacists

None.

18.6.11 Copayment for Patients

There are five categories of reimbursement (A, B, C, Cx, and Cs) that define the level of copayment for the patients. The actual categorization is a function of the severity of the diseases with drugs in category A for life-threatening diseases, B for therapeutically significant drugs for nonlife-threatening diseases (e.g., antibiotics, antiasthmatics, and antihypertensives), and drugs in category C for symptomatic treatments.[1]

Patients are required to cover the full cost of the drug if they prefer a more expensive version of a drug and refuse any of the three cheapest products for drugs prescribed by INN.[2] However, doctors can add a "no switch" notice to the prescription, in which case this rule does not apply.

18.6.12 Special Funding Procedure for Individual Patients

Special Solidarity Fund exists to help individual patients in a very serious medical condition obtaining essential medical services that are not reimbursed and particularly expensive.[7] In can also provide an additional safety net to cover "regular" medical care insurance.

Request is done by an individual patient, so this policy has no influence of the pharmaceutical company. College of Medical Directors decides whether to grant allowances and determines their amount.

REFERENCES

1. le Polain M, Franken M, Koopmanschap M, Cleemput I. Drug reimbursement systems: International comparison and policy recommendations. Health Services Research (HSR). Brussels, Belgium: Belgian Health Care Knowledge Centre (KCE). 2010. KCE Reports 147C. D/2010/10.273/90.
2. IMS Pharmaceutical Pricing & Reimbursement Concise Guide, Belgium, 2015.
3. Davies JE, Neidle S, Taylor DG. Developing and paying for medicines for orphan indications in oncology: Utilitarian regulation vs equitable care? *Br J Cancer* 106(1):14–17, 2012.
4. Maervoet J, Toumi M. PHP132 Time to Market Access for Innovative Drugs in the UK, France, and Belgium. *Value Health* 15(7):A312, 2012.
5. Remuzat C et al. Overview of external reference pricing systems in Europe. *J Mark Access Health Policy* 3, 2015. doi: 10.3402/jmahp.v3.27675.
6. Adriaen M, De Witte K, Simoens S. Pricing Strategies of Originator and Generic Medicines following Patent Expiry in Belgium. *J Generic Med* 5:175–187, 2008.
7. Adriaens C, Van De Vijver I. Pricing & Reimbursement of medicines in Belgium. COOPAMI. Oct 2015.

19 The United States

19.1 STAKEHOLDERS

19.1.1 NAMES OF NATIONAL PRICING AND REIMBURSEMENT DECISION MAKERS

The federal government (Medicare and Medicaid programmes) and private insurance providers are key decision makers in the United States. The Centers for Medicare and Medicaid Services (CMS) oversee the implementation and financing of Medicare and Medicaid programmes.[1,2]

The Food and Drug Administration (FDA) grants marketing authorization in the United States for original drugs, generics, and over-the-counter (OTC) drugs. The FDA is located within the U.S. Department of Health and Human Services.[3]

19.1.2 NAMES OF NATIONAL HEALTH TECHNOLOGY ASSESSMENT AGENCIES

There are several health technology assessment (HTA) agencies and entities that either conduct evidence-based assessments or grade the evidence levels for interventions using HTA:

- The Agency for Healthcare Research and Quality's (AHRQ) programmes conduct systematic evidence reviews to assess the effectiveness, comparative effectiveness, safety, and, in rare instances, the cost-effectiveness of medical technologies and interventions.[4]
- The Medicare Evidence Development and Coverage Advisory Committee (MEDCAC) provides an independent evaluation of evidence, including HTA, public testimony, and all other data that may be used to illustrate the benefits or risks of a given product under consideration for coverage under Medicare and is in charge of weighing the evidence from the HTA.[5]
- Drug Effectiveness Review Project (DERP) is a collaborative programme of state's Medicaid and public pharmacies dedicated to producing comparative, evidence-based research that assists policymakers and other decision-makers.[6]
- The Department of Veterans Affairs' (VA) Pharmacy Benefits Management Strategic Healthcare Group (PBMSHG) undertakes pharmaceutical technology assessments to support the appropriate use of medications within the VA health care system.[7]
- The Department of Defense Pharmacoeconomic Center (PEC) undertakes cost-effectiveness studies of existing and new pharmaceuticals to support the decision-making processes.[8]

19.1.3 NAMES OF OTHER KEY STAKEHOLDERS (REGIONAL/LOCAL LEVEL)

Each of the 50 states is in charge of implementing the federal programmes according to local need.

19.2 OVERVIEW OF PRICING AND REIMBURSEMENT POLICIES

19.2.1 OVERVIEW OF THE SYSTEM

Health care is provided via the federal programmes such as Medicare and Medicaid, military (TRICARE) or Veterans Administration (VA) programmes, and private health plans.[9] Medicare covers U.S. citizens 65 years or older and people under the age of 65 years with certain disabilities and patients suffering from end-stage renal disease.[10] The programme is composed of four parts: Parts A, B, C, and D. Parts A and B, also known as original Medicare, are administered by the federal government. Part A covers inpatient care, such as hospital and nursing care, while Part B covers ambulatory and preventative care.[9] Part D, also known as the outpatient prescription drug insurance, is provided by private insurance companies and covers outpatient prescription drugs.[9] In Part C, or Medicare advantage, Parts A and B benefits are provided by a private insurer that contracts with Medicare to provide coverage.[9]

The Medicaid and the Children's Health Insurance Programme (CHIP) is a joint federal state initiative that covers low-income U.S. citizens or families, legal residents, people with disabilities, and elderly individuals needing nursing home care.[9]

Private health plans can be employer-provided and are divided into three main categories: managed care plans, where beneficiaries have to seek care from an approved network of physicians and hospitals; fee-for-service, which offers coverage for a range of pre-specified medical services; and high-deductible health plans.

Military coverage is available through the VA and TRICARE federal programs. Individuals who served in the active military service in good standing may have coverage under VA health care benefits (http://www.va.gov/HEALTHBENEFITS/apply/veterans.asp). In addition to VA benefits, TRICARE is a regionally managed federal health-care program that is available for active duty and retired members of the uniformed services, their families, and survivors (http://www.tricare.mil/Plans/Eligibility.aspx).

Once the marketing authorization is granted by the FDA, manufacturers are free to set the price of the pharmaceuticals. The reimbursement of drugs is set by private payers and federal programmes (e.g., Medicaid and Medicare) by using a range of different criteria.

19.2.2 Reimbursement Process

There is no central reimbursement policy in the United States. Reimbursement is decided at federal, state, and private-payer levels by using separate and distinct decision criteria.

Private payers use formularies to decide on reimbursement. Formularies are managed by the private payers' pharmacy and therapeutic committees using evidence-based assessments, considering the cost of a given drug and the expected use of the drug. Most OTCs, lifestyle drugs, and experimental drugs are excluded from these formularies.

For Medicare, while all the hospital inpatient drugs approved by the FDA and included in U.S. Pharmacopeia (USP) are fully reimbursed by Medicare Part A, only some outpatient drugs, such as drugs requiring administration by a physician and oral anti-cancer drugs, are covered by Part B. In Part D, outpatient prescription drugs are reimbursed by private health insurers if the drugs meet all the criteria for reimbursement. The drug must be an FDA-approved prescription-only medicine (POM), not covered by Parts A and B and not excluded from coverage in Part D. Manufacturers must also sign an agreement with the CMS to cover 50% of the cost of drugs.

The states provide coverage for all FDA-approved POMs to the Medicaid beneficiaries. Manufacturers who seek reimbursement via Medicaid must agree to rebates, enroll in the federal 340B prescription drug programme—a programme that allows health care institutions (14,500 approved clinics, hospitals, and other entities) to purchase drugs at discounted price to dispense to low-income patients—and have their products listed on the federal supply schedule.[11]

19.2.3 Pricing Process

Prices of prescription medicines, hospital drugs, generics, and OTC drugs are not controlled, except for military institutions.[9] Manufacturers are free to set their price after they have been granted marketing authorization. The federal government, states, and private payers must secure rebates and discounts to regulate pharmaceutical prices. Thus, several pricing benchmarks, such as the average manufacturer price (AMP), the average sale price (ASP), the average wholesale price (AWP), the federal upper limit (FUL), the maximum allowable cost (MAC), the Medicaid best price, and the wholesale acquisition cost (WAC), have been created over the years to meet the needs of all payers and help determine discounts, rebates, and reimbursement prices.[9,12]

19.3 TIME TO MARKET

Time to market is immediate on receiving marketing authorization, since prices are not regulated.

19.4 PRICE REGULATIONS

19.4.1 Pricing Policy Following the Marketing Authorization

Manufacturers are free to set the price of their product on approval by the FDA.

19.4.2 External Reference Pricing

External reference pricing is not used in the United States.

19.4.3 Internal Reference Pricing

Internal reference pricing is not used at federal level in the United States. It is used infrequently by health care programmes.

19.4.4 Price Control at Ex-Factory Price Level

There is no price control at ex-factory price level.

19.4.5 Price Control at Wholesale Level

There is no price control at wholesale level, and wholesaler margins are not fixed.

19.4.6 Price Control at Pharmacy Retail Level

There is no price control at pharmacy retail level.

19.4.7 Mandatory Price Reduction on Brand Price after Generic/Biosimilar Entry

Price reduction on brand price is not mandatory after generic or biosimilar entry.

19.5 REIMBURSEMENT SPECIFICITIES

Reimbursement varies from one programme to another. Indeed, different types of formularies can be used by payers, such as open formularies, closed formularies, mixed formularies with closed and open lists, and generics-only formularies. While open formularies include new drugs as soon as they are granted marketing authorization, a new drug must be approved by the private payer's pharmacy and therapeutic committees before inclusion in a closed formulary.

There are no reimbursement categories in Medicare Parts A and B and Medicaid: drugs are either reimbursed or not, but formularies are used in Medicare Part D. These formularies are specific to Part D owing to the restrictions imposed by CMS. These formularies must include all or substantially all drugs in six therapeutic classes (immunosuppressant, antidepressant, antipsychotic, anticonvulsant, antiretroviral, and antineoplastic classes), as well as at least two drugs from all other therapeutic categories.[13]

19.6 CHARACTERISTICS OF PUBLIC TENDERING

Public tendering is not applied in the United States. However, states can participate in multi-state Medicaid purchasing pools and secure supplementary rebates through this purchasing tool.

19.7 EXPENDITURE CONTROLS

19.7.1 Discounts/Rebates

Discounts and rebates are widely used by the states and private payers to regulate pharmaceuticals' prices.[9] Private payers often seek discounts from the manufacturers in return for the drug inclusion in the formularies.[9] These discounts are calculated as a percentage of the AWP, with the average discount being about 17.5% of the AWP in 2010. Hospitals can negotiate for Parts A and B, and private insurers can negotiate for Part D. However, Medicare is not allowed by law to obtain discounts for its Part D.

Manufacturers are required to provide a fixed rebate for Medicaid-covered prescription drugs. The fixed rebate is calculated based on the AMP and the best price. States can also secure supplementary rebates via one of the three multi-state Medicaid purchasing pools.[14]

19.7.2 Clawback

States are required to reimburse the federal government a percentage (beginning at 90% and declining to 75% by 2015) of the cost that the states would have spent if dual-eligible (Medicare-Medicaid) patients were covered by Medicaid instead of Medicare Part D.[15] This system of states returning funds to the federal government is referred to as a *clawback*.

19.7.3 Payback

Paybacks are applied to the manufacturers of branded pharmaceuticals, generics, and biosimilars. This payback is applied in the form of a user fee to the federal government to fund the marketing authorization process (Prescription Drug User Fee Act) and to help assure the timely review of all abbreviated new drug applications. Manufacturers of branded pharmaceuticals have been requested to pay annual paybacks since 2010. The amount of this fee depends on the manufacturer's market share of branded pharmaceuticals sold to Medicare Parts B and D, Medicaid, the Veteran Administration programme, and the Department of Defense beneficiaries.

19.7.4 Price-Volume Agreements

Price-volume agreements are not applied in the United States; instead, expenditure caps are applied to control the pharmaceutical expenditure.[16]

19.7.5 Other Market Access Agreements

Several market access agreements are used in the United States such as coverage with evidence development, conditional treatment continuation, and performance-linked reimbursement; however, the number of risk-sharing agreements remains limited in the United States.[17]

19.7.6 Price Freezes and Cuts

Price freezes and cuts are not applied in the United States, since pharmaceuticals' prices are not regulated.

19.8 POLICIES TARGETED AT WHOLESALERS, PHARMACISTS, PHYSICIANS, AND PATIENTS

19.8.1 Wholesaler Mark-Up

Wholesaler mark-ups are estimated as "modest" (between 2% and 4%) and have declined over the years.[16,18,19]

19.8.2 Pharmacy Mark-Up

Pharmacy mark-ups are usually composed of a fixed cost plus a part that varies depending on the drug. These mark-ups are estimated between 20% and 25% over the pharmacy's acquisition price.[20]

19.8.3 Generic Substitution

While generic substitution is allowed in all states, it is only mandatory in some. Generic substitution is mandatory in 15 states, while 35 states have permissive substitution law (2011).[21] Pharmacists are required to substitute brands with generics in 13 states. In some states, pharmacists are even required to substitute with the cheapest pharmaceutical.[22] Generic substitution is allowed under Medicare Part D and Medicaid. Pharmaceuticals with therapeutic equivalence evaluations are listed in the Orange Book drawn by the FDA; however, generic substitution is not regulated by the FDA.[23]

19.8.4 International Nonproprietary Name (INN) Prescribing

The INN prescribing is allowed in the United States but is not mandatory.

19.8.5 Prescription Guidelines

Guidelines are drawn by several organizations, such as the Federation of State Medical Boards, the State Medical Boards for each state, and the Agency for Healthcare Research and Quality, as well as by medical colleges, hospitals, and health plans. These guidelines provide information on best medical practice and can be found in the National Guideline Clearinghouse database.[24]

Controlled substances, such as pain medicines prescriptions, are regulated by guidelines at the federal and state levels, with the most stringent law applied when federal and state laws differ.

19.8.6 Monitoring of Prescribing Behavior

Prescribing behavior of physicians is monitored by states and private payers either directly via e-prescription systems or by reviewing the prescription records on a regular basis.

19.8.7 Pharmaceutical Budgets Defined for Physicians

Budgets are not defined for physicians in private and federal programmes.

19.8.8 Prescription Quotas

Prescription quotas are not used in the United States.

19.8.9 Financial Incentives for Physicians

Physicians do not have financial incentives to prescribe generics or cheapest drugs.[25] However, physicians' compensation can be either fixed or variable, and variable compensations are tied more to the productivity, with incentives for performance (68%), and less to patient satisfaction (21%), care quality (19%), or resource use (14%).[25] Incentives for performance, also known as pay-for-performance programmes, are growing rapidly and have been implemented by governmental (Medicare) and private programmes (California Pay for Performance Program).

19.8.10 Financial Incentives for Pharmacists

Pharmacists have indirect incentives to dispense the cheapest generics, since public and private payers reimburse pharmacies a fixed amount, regardless of which equivalent pharmaceutical is dispensed. Thus, providing the cheapest generics allows pharmacists to increase their margins.[26]

19.8.11 Copayment for Patients

Copayments for patients are used by public and private health insurance programmes. Most private health insurance uses multiple-tiered formularies with different levels of copayment. While the first tier is usually limited to generic and low-cost drugs, the second and third tiers are composed of more expensive drugs and branded pharmaceuticals and the fourth is used for specialty drugs for complex diseases. The number of tiers and the level of copayment by tiers vary from one private health plan to another and from one drug to another.

The Medicaid copayment are decided on state level, but the maximum copayment level is fixed at federal level and cannot be exceeded by states. As for the Medicare Part A, patients are required to pay a deductible of $1,216 for each benefit period and co-pay the hospital stay and the nursing facility stay after the 61st day and 21st day, respectively.[27] Patients covered by Part B must pay an annual deductible and a monthly premium, which varies by plan. They must also pay a percentage of the cost of care.[27] This equates to a deductible of $147 each year, a monthly premium that depends on income, and an additional 20% out-of-pocket expense of all approved costs beyond the deductible.[27]

REFERENCES

1. The Center for Medicare and Medicaid Services website: Available from: http://www.cms.gov/ (accessed August 10, 2016).
2. Sullivan SD, Watkins J, Sweet B, Ramsey SD. Health Technology Assessment in Health-Care Decisions in the United States. *Value in Health*. 2009;12(S2):S39–S44.
3. The Food and Drug Administration website: Available from: http://www.fda.gov/default.htm (accessed August 10, 2016).
4. Agency for Healthcare Research and Quality website: Available from: http://www.ahrq.gov/ (accessed August 10, 2016).
5. The Center for Medicare and Medicaid Services. Regulations-and-Guidance. The Center for Medicare and Medicaid Services website: Available from: http://www.cms.gov/Regulations-and-Guidance/Guidance/FACA/MEDCAC.html (accessed August 10, 2016).
6. Drug Effectiveness Review Project. Oregon Health & Science University website: Available from: http://www.ohsu.edu/xd/research/centers-institutes/evidence-based-policy-center/derp/index.cfm (accessed August 10, 2016).
7. The Department of Veterans Affairs website: Available from: http://www.va.gov/health/ (accessed August 10, 2016).
8. The Department of Defense Pharmacoeconomic Center website: Available from: http://www.pec.ha.osd.mil/formulary_search.php (accessed August 10, 2016).
9. Schoonveld E. *The Price of Global Health: Drug Pricing Strategies to Balance Patient Access and the Funding of Innovation*. Gower Publishing, Surrey, UK, 2011.
10. Komisar HL, Feder J, Gilden D. The Commonwealth Fund. The Roles of Medicare and Medicaid in Financing Health and Long-Term Care for Low-Income Seniors. Sep 2000.
11. Conti RM, Bach PB. Cost Consequences of the 340B Drug Discount Programme. *JAMA*. 2013;309(19):1995–1996.
12. Curtiss FR, Lettrich P, Fairman KA. What is the price benchmark to replace average wholesale price (AWP)? *J Manag Care Pharm*. 2010;16(7):492–501.
13. The Center for Medicare and Medicaid Services. Chapter 6: Part D Drugs and Formulary Requirements. Medicare Prescription Drug Benefit Manual. 2010. Available from: http://www.cms.gov/Medicare/Prescription-Drug-Coverage/PrescriptionDrugCovContra/downloads/chapter6.pdf (accessed August 10, 2016).
14. Medicaid Drug Rebate Program Available from: http://www.medicaid.gov/Medicaid-CHIP-Program-Information/By-Topics/Benefits/Prescription-Drugs/Medicaid-Drug-Rebate-Program.html (2014) (accessed August 10, 2016).
15. Kaiser Commission on Medicaid and the Uninsured. An Update on the Clawback: Revised Health Spending Data Change State Financial Obligations for the New Medicare Drug Benefit. Available from: http://www.kff.org/medicaid/upload/7481.pdf (2006) (accessed August 10, 2016).
16. Ando G. Roadblocks and Risk-Sharing Agreements Around the World. *ISPOR Connection*. 2011.
17. Neumann PJ, Chambers JD, Simon F, Meckley LM. Risk-Sharing Arrangements That Link Payment For Drugs To Health Outcomes Are Proving Hard To Implement. *Health Aff*. 2011;30:122329–2337.
18. The Food and Drug Administration. Regulatory Information. Last updated 25 May 2011. Available from: http://www.fda.gov/RegulatoryInformation/Legislation/FederalFoodDrugandCosmeticActFDCAct/SignificantAmendmentstotheFDCAct/PrescriptionDrugMarketingActof1987/ucm256477.htm#foot1 (accessed August 10, 2016).
19. Berndt ER. Pricing and Reimbursement in U.S. Pharmaceutical Markets. Harvard School of Public Health-Faculty Research Working Paper Series. Sep 2010.
20. The Department of Health & Human Services. *Report to the President—Prescription Drug Coverage, Spending, Utilization, and Prices*. Apr 2000.
21. Pharmaceutical Care Management Association. Generic Substitution: The Science and Savings. 2011. Available from: http://amcp.org/WorkArea/DownloadAsset.aspx?id=10530 (accessed August 10, 2016).

22. National Conference of State Legislatures. Health Cost Containment and Efficiencies. NCSL Briefs for State Legislators. 2010.

23. Approved Drug Products with Therapeutic Equivalence Evaluations (Orange Book). FDA webite: Available from: http://www.fda.gov/Drugs/InformationOnDrugs/ucm129662.htm (accessed August 10, 2016).

24. The National Guideline Clearinghouse database website: Available from: http://guideline.gov (accessed August 10, 2016).

25. Chien AT, Chin MH, Alexander CG, Tang H, Peek ME. Physician Financial Incentives and Care for the Underserved in the United States. *The Am. J.Manag. Care*. Feb 2014.

26. Danson PM, Furukawa MF. Cross-National Evidence on generic pharmaceuticals: Pharmacy vs. physician-driven markets. National Bureau of Economic Research Working Paper 17226. Jul 2011. Available from: http://www.nber.org/papers/w17226 (accessed August 10, 2016).

27. Medicare 2014 Costs At a Glance Available from: http://www.medicare.gov/your-medicare-costs/costs-at-a-glance/costs-at-glance.html#collapse-4810 (accessed August 10, 2016).

20 Japan

20.1 STAKEHOLDERS

20.1.1 NATIONAL PRICING AND REIMBURSEMENT DECISION MAKERS

The Ministry of Health, Labour and Welfare (MHLW) sets the National Health Insurance (NHI) reimbursement price after evaluation of a request for inclusion of new drug into the price list provided by the manufacturer.[1]

Several bureaus of the MHLW are involved in the submission and evaluation of the combined pricing and reimbursement application. The Economic Affairs Division of the Health Policy Bureau conducts a formal hearing, where manufacturers can submit their application, and the Medical Economics Division of the bureau reviews the application before forwarding it to the Drug Pricing Organization.[1]

The Drug Pricing Organization (DPO) evaluates the manufacturer's application and then the Central Social Insurance Medical Council (Chuikyo), which is the decision body for the medical service fee, including drug price, eventually approves the pricing after discussion among payers, health care providers, and public representatives.

20.1.2 NATIONAL HEALTH TECHNOLOGY ASSESSMENT AGENCIES

There are no national health technology assessment agencies; the MHLW evaluates the manufacturer's application and sets the price and reimbursement.

20.1.3 OTHER KEY STAKEHOLDERS (REGIONAL/LOCAL LEVEL)

There are no stakeholders at regional and local levels.

20.2 OVERVIEW OF PRICING AND REIMBURSEMENT POLICIES

20.2.1 OVERVIEW OF THE SYSTEM

Pricing and reimbursement are sought through a combined pricing and reimbursement application provided by the manufacturer, and only reimbursed drugs are subject to price controls.[2] The application can be submitted to the MHLW only after a marketing authorization has been granted. The application provided by the manufacturer must contain data about the desired price; the price of the drug in other global markets, in particular, in France, Germany, the United Kingdom, and the United States; the anticipated patient population; and sales forecasts for first 10 years in the market. The MHLW sets the NHI reimbursement price after evaluation of the application.

20.2.2 REIMBURSEMENT PROCESS

The MHLW decides on the reimbursement of each drug based on the recommendation of the Central Social Insurance Medical Council (Chuikyo).

The reimbursement is applicable once a new drug is added in the NHI reimbursement price list. There are several levels of reimbursement, depending on the age and income.[3] At least 70% of actual cost is covered by the program.[4]

20.2.3 PRICING PROCESS

The main division implicated in the application evaluation and price setting is the DPO. The DPO is composed of doctors, dentists, pharmacists, and economists. The DPO members evaluate the application and decide on a proposed price by a majority vote, which is eventually approved by Chuikyo. The NHI reimbursement price is then added to the NHI reimbursement price list at the next update (four updates a year).

The manufacturer can object to the DPO's decision. This leads to a re-evaluation of the submission by the DPO, and if no agreement is found, the manufacturer can withdraw its submission to resubmit a new application.

The DPO members evaluate each new drug application by taking into account the presence of similar drugs in the market and of comparators, the applicability of price premiums, and the costs. One of the pricing criteria is the innovation level of the drug, and premium can be added depending on additional benefits such as novelty, utility, marketability, and pediatric use.

20.3 TIME TO MARKET

A lag exists between marketing authorization and launch, since manufacturer can submit a pricing and reimbursement application only after obtaining a marketing authorization. There is usually a delay of two or three months between marketing authorization and inclusion in the NHI reimbursement list, and sales between those two events are estimated as negligible.

20.4 PRICE REGULATIONS

20.4.1 PRICING POLICY FOLLOWING THE MARKETING AUTHORIZATION

Following marketing authorization, manufacturers are free to set their price. Only reimbursed drugs' prices are controlled, and manufacturer seeking reimbursement must submit an application to the MHLW.

20.4.2 External Reference Pricing

The NHI reimbursement price is compared with the public price of four countries: France, Germany, the United Kingdom, and the United States. These countries were chosen because their market is estimated as comparable to the Japanese market. The NHI reimbursement price can be adjusted depending on the average price of these four reference countries. Indeed, the NHI reimbursement price can be increased if it equals 75% or less of the average price of the reference countries or be decreased if it equals 125% or more.

20.4.3 Internal Reference Pricing

Two methods are used to define the new drug's price: the comparative price method and the cost calculation method. The comparative price method is used when a comparator is available on the Japanese market, and the cost calculation method is used when no active comparator is found on the market.

For the comparative price method, the comparator is defined as a similar drug in the aspects of indications, effect, pharmacological action, composition and chemical structural formula, route of administration, dosage form, formulation category, formulation, and administration. A drug price can be granted premiums, for usefulness, innovation, limited marketability, pediatric use, and first in class, which is the premium for drug first approved in Japan globally, if it is more effective than the comparator or if it is used in pediatrics or for the treatment of a rare disease. The premiums vary between +5% and +120% but are rarely applied. When a drug is less effective than the comparator and does not belong to the previous cited classes, then its price is decided using the price of the same class of drugs (average prices or lowest price).

For the cost calculation method, the calculated price takes into account the cost incurred per unit for manufacture and marketing, the selling costs, the operating profits, distribution costs, and consumption tax.

20.4.4 Price Control at Ex-Factory Price Level

Only prescribed reimbursed drugs are subjected to price control.

20.4.5 Price Control at Wholesale Level

There is no price control at wholesale level, since the wholesale margin is not fixed but rather negotiated (discounts on NHI reimbursement prices).

20.4.6 Price Control at Pharmacy Retail Level

There is no price control at pharmacy retail level, since the pharmacy margin is not fixed; it is negotiated between wholesalers and manufacturers (discounts).

20.4.7 Mandatory Price Reduction on Brand Price after Generic/Biosimilar Entry

Branded drug price is reduced by 4%–6% after the entry of the first generic in the market.

20.5 REIMBURSEMENT SPECIFICS

Only drugs that are on the NHI reimbursement list are reimbursed. This list is a list of prescribed drugs, including branded drugs, generics, and hospital drugs.

20.6 CHARACTERISTICS OF PUBLIC TENDERING

Tenders are rare in Japan and paralel negotiation with multiple suppliers is a more common approach. However, tenders may occur between the drug wholesellers and hospitals. Therefore, manufactuers deal directly with wholesellers rather than with hospitals.

20.7 EXPENDITURE CONTROLS

20.7.1 Discounts/Rebates

Discount and rebates occur only between manufacturers and wholesalers and between wholesalers and pharmacists.

20.7.2 Clawback

There are no clawbacks.

20.7.3 Payback

There are no paybacks.

20.7.4 Price-Volume Agreements

There are no price-volume agreements. However, when the request for inclusion of new drug into the price list is submitted, the manufacturer is obliged to show the 10-year prediction of number of patients and the sales. If the real sales are more than twice the expected sales and ¥15 billion, the price is reduced up to 25%.

20.7.5 Other Market Access Agreements

Other market access agreements and other kinds of agreements are not used in Japan.[4]

20.7.6 Price Freezes and Cuts

Price cuts are used in Japan to control health care expenditure. Cuts are decided by the Central Social Insurance Medical Council and implemented by the Ministry of Health, Labour and Welfare. They take place with the NHI reimbursement price revision every 2 years. The reimbursement price of a drug can be reduced if the actual market price is low compared with the NHI reimbursement price or if the drug's sales are much higher than the expected sales.

20.8 POLICIES TARGETED AT WHOLESALERS, PHARMACISTS, PHYSICIANS, AND PATIENTS

20.8.1 Wholesaler and Pharmacy Mark-Up

Wholesaler and pharmacy mark-ups are negotiated.

20.8.2 Generic Substitution

Generic substitution is allowed since 2006, and physicians have to tick a box each time they want to prohibit a single-drug substitution.

20.8.3 International Nonproprietary Name (INN) Prescribing

Physicians are encouraged to write prescriptions by INN by gaining additional ¥20 fee for INN use.

20.8.4 Prescription Guidelines

There are no guidelines.

20.8.5 Monitoring of Prescribing Behavior

There is no monitoring of prescribing behavior.

20.8.6 Pharmaceutical Budgets Defined for Physicians

There is no budget defined for physicians.

20.8.7 Prescription Quotas

There are no prescription quotas.

20.8.8 Financial Incentives for Physicians

There are financial incentives for prescribing fewer drugs. Indeed, physicians are rewarded when they prescribe fewer drugs per visit. Physicians are also rewarded when the drug is dispensed by a pharmacist and not by the physician himself.

20.8.9 Financial Incentives for Pharmacists

Pharmacists are encouraged to distribute generics: they are awarded by a premium when they distribute 22% or more generics.

20.8.10 Copayment for Patients

Patients have to pay fixed copayments, depending on their age and income (between 0% and 30% of their medical care). In addition to the basic medical insurance coverage by the NHI, the high-cost health care benefit system covers a certain amount of medical expense if the medical cost exceeds the set-up price.[5]

REFERENCES

1. Ministry of Health, Labour and Welfare Available from: http://www.mhlw.go.jp/english/(accessed August 10, 2016).
2. English Regulatory Information Task Force-Japan Pharmaceutical Manufacturers Association. Pharmaceutical Administration and Regulations in Japan. 2012. Available from: http://www.jpma.or.jp/english/parj/pdf/2012.pdf (accessed August 10, 2016).
3. Pharmaceutical Administration and Regulations in Japan, Japan Pharmaceutical Manufacturing Association, 2015. Available from: http://www.jpma.or.jp/english/parj/pdf/2015.pdf (accessed August 10, 2016).
4. Ando G. Payer Roadblocks and Risk-Sharing Agreements Around the World. Ispor 2011. Available from: http://www.ispor.org/news/articles/Nov-Dec11/Payer-Roadblocks-and-Risk-Sharing-Agreements.asp (accessed August 10, 2016).
5. Outline of Health Care Insurance System, Ministry of Health, Labour, and Welfare, 2009. Available from: http://www.mhlw.go.jp/english/wp/wp-hw3/dl/2-001.pdf (accessed August 10, 2016).

21 China

21.1 STAKEHOLDERS

21.1.1 NAMES OF NATIONAL PRICING AND REIMBURSEMENT DECISION MAKERS

The National Health and Family Planning Commission (NHFPC), formerly known as the Ministry of Health, decides on health care policies and provides financing for the reimbursement.[1,2]

The Ministry of Human Resources and Social Security (MOHRSS) decides on health insurance policies and on the inclusion of prescribed drugs on the National Reimbursement Drug List.[3]

The National Development and Reform Commission (NDRC) sets the prices of reimbursed prescription drugs.[4]

21.1.2 NAMES OF NATIONAL HEALTH TECHNOLOGY ASSESSMENT AGENCIES

There is no national health technology assessment agency in China.

21.1.3 NAMES OF OTHER KEY STAKEHOLDERS (REGIONAL/LOCAL LEVEL)

National health policies are implemented by regional and local divisions in the 5 autonomous regions, 4 municipalities, and 23 provinces.

21.2 PRICING AND REIMBURSEMENT POLICIES

21.2.1 OVERVIEW OF THE SYSTEM

Manufacturers must submit an application to the NDRC, even for non-reimbursed medicines.

Until June 1, 2015, the prices of only those prescription drugs, generics, and over-the-counter drugs that were on the National Reimbursement Drug List or the Essential Drugs List or were supplied by government programmes were controlled. Other drugs were, and continue to be, free-priced.

The prices of prescription drugs listed on the National Reimbursement Drug List or supplied by government programmes were set by the NDRC as "government pricing" (fixed manufacturer prices) or "government-guided pricing" (fixed maximum retail price). These regulations were replaced by the new law introduced on June 1, 2015 (see Section 8.11).

The National Reimbursement Drug List is decided at a national level by the MOHRSS, and each province decides on its own reimbursement list by using the National Reimbursement Drug List as a reference. The reimbursement lists consist of western drugs and traditional Chinese medicines.

21.2.2 REIMBURSEMENT PROCESS

Drugs that are approved by the State Food and Drug Administration are included by the MOHRSS on the reimbursement list depending on their clinical use, efficacy and safety, price, administration mode, and availability of the suppliers.

The reimbursement list is updated in theory every 2 years by reviewing drugs listed on the regional reimbursement list. The list is reviewed by a board of experts, which issues recommendation and compiles a new list. The new list is reviewed by another board of experts, which votes on the inclusion of new drugs. The list goes back to the first board of experts, which reviews it and forwards it to the reimbursement committee that has to validate the list before its publication by the MOHRSS.

21.2.3 PRICING PROCESS

Until June 1, 2015, the NDRC controlled the maximum retail price of drugs listed on the national reimbursement list by restricting the ex-factory price and setting the maximum distribution margins.

The ex-factory price was set based on the level of innovation of the product, the production costs, before-sale costs, and sales profit. The prices were set mostly by using the principle of "average social cost," that is, the average cost of producing the drug, based on the established production costs in the marketplace. However, manufacturer could apply for a differential pricing (higher price) for off-patent drugs, provided that a significant superiority of the drugs in terms of safety, quality, and efficacy could be proved compared with equivalent drugs.

The NDRC also controlled the ex-factory price of drugs used in government programmes, such as contraceptives and vaccines. In this case, the ex-factory price was set based only on the production costs.

Provinces controlled the prices of drugs that were listed only on regional reimbursement lists or that did not have a price set at national level. These regulations were replaced by the new law introduced on June 1, 2015 (see Section 8.11).

21.3 TIME TO MARKET

Time to market is long in China. It takes around 5 years for a drug to get market authorization and 2 years to be listed on the national or regional reimbursement lists.

21.4 PRICE REGULATIONS

21.4.1 PRICING POLICY FOLLOWING THE MARKETING AUTHORIZATION

Manufacturers must file a pricing application with the NDRC for all drugs, even for the free-priced ones. However, until June 1, 2015, the NDRC controlled prices of drugs that were on the national reimbursement list.

Provinces could set the price of drugs that were on regional reimbursement list and that were not listed at the national level under the guidance of the NDRC. When no price had been set at the national or regional level, manufacturer could set a temporary price until the publication of a national or regional price. These regulations were replaced by the new law introduced on June 1, 2015 (see Section 8.11).

21.4.2 EXTERNAL REFERENCE PRICING

China does not use an external reference pricing system. However, manufacturers are required to provide the prices in other Asian and even European countries for imported products.

21.4.3 INTERNAL REFERENCE PRICING

An internal reference pricing is planned in the upcoming reform; however, owing to concerns over varying quality of manufacturing and the fragmentation of the pharmaceutical market in China, internal reference pricing is unlikely to be used soon.

21.4.4 PRICE CONTROL AT EX-FACTORY PRICE LEVEL

Until June 1, 2015, the ex-factory price of reimbursed drugs was set based on the product innovation level and the production costs. Surveys were carried out to monitor the ex-factory price of drugs on the national reimbursement list. These regulations were replaced by the new law introduced on June 1, 2015 (see Section 8.11). Prices of drugs that are not listed on national or regional reimbursement lists were, and continue to be, not controlled.

21.4.5 PRICE CONTROL AT WHOLESALE LEVEL

Until June 1, 2015, the NDRC sets the maximum distribution margins for wholesalers and retailers for drugs listed on national and regional reimbursement lists. Surveys were also carried out at the wholesale level to monitor the volumes of sales and the prices. These regulations were replaced by the new law introduced on June 1, 2015 (see Section 8.11).

21.4.6 PRICE CONTROL AT PHARMACY RETAIL LEVEL

The maximum retail price of drugs listed on the national reimbursement list is set by the NDRC. Pharmacies are monitored through surveys, and they must provide the actual retail prices, sale volumes, and revenues.

21.4.7 MANDATORY PRICE REDUCTION OF PRICE OF ON-PATENT DRUGS AFTER GENERIC/BIOSIMILAR ENTRY

There are no mandatory price reductions of price of on-patent drugs after generic or biosimilar entry.

21.5 REIMBURSEMENT SPECIFICITIES

There are two classes of reimbursed drugs, A and B. Class A is for drugs listed on the Essential Drug List, necessary drugs, and highly used drugs. Class B is for drugs that have higher prices than Class A drugs. Class A drugs are fully reimbursed, and reimbursement level of Class B drugs is decided at the regional level.

21.6 CHARACTERISTICS OF PUBLIC TENDERING

Competitive tendering is applied only to hospitals since 2001. The retail price for hospital drugs is set at the price of the winning bid plus a distribution margin. In China, hospitals are the main supplier of pharmaceutical products. It is estimated that only approximately 18% of all prescription drugs (by volume) are sold via retail pharmacies.

21.7 EXPENDITURE CONTROLS

21.7.1 DISCOUNTS/REBATES

There are no mandatory discounts or rebates.

21.7.2 CLAWBACK

There are no clawbacks.

21.7.3 PAYBACK

There are no paybacks in place.

21.7.4 PRICE-VOLUME AGREEMENTS

There are no price-volume agreements.

21.7.5 OTHER MARKET ACCESS AGREEMENTS

There are no such schemes in China.

21.7.6 PRICE FREEZES AND CUTS

Unlit June 1, 2015, the NDRC adjusted the price of drugs by using price cuts (approximately twice a year). Cuts varied from

15% to 30% and targeted several therapeutic areas (31 rounds of mandatory price cuts on 1000 drugs from 1998 to 2013). It is to be seen how it will be affected by the new law introduced on June 1, 2015 (see Section 8.11).

21.8 POLICIES TARGETED AT WHOLESALERS, PHARMACISTS, PHYSICIANS, AND PATIENTS

21.8.1 WHOLESALER AND PHARMACY MARK-UP

Maximum distribution mark-ups are applied at both wholesaler and retailer levels.

21.8.2 GENERIC SUBSTITUTION

Pharmacists are not allowed to substitute one product for another if the prescription is by brand name, and patients can choose a specific brand if the prescription is by active substance.

21.8.3 INTERNATIONAL NONPROPRIETARY NAME (INN) PRESCRIBING

Physicians can use INN or brand name when prescribing a drug.

21.8.4 PRESCRIPTION GUIDELINES

Guidelines have been published since 2010 to promote rational use of medicines and better clinical practice.

21.8.5 MONITORING OF PRESCRIBING BEHAVIOR

Provinces and hospitals monitor doctor's prescription, and the information is used to advise on prescription behaviors. Physicians that prescribe very expensive drugs or have an excessive or unreasonable prescribing behavior can be fined and may even lose their license for inappropriate prescribing behavior.

21.8.6 PHARMACEUTICAL BUDGETS DEFINED FOR PHYSICIANS

The NHFPC controls the income that hospitals can derive from drug sales in proportion of the total income. Hospitals that exceed the cap fixed by the NHFPC have to pay a penalty, which is usually passed to the doctor who exceeded his prescription limit. Physicians are required to justify the prescription of costly drugs.

21.8.7 PRESCRIPTION QUOTAS

There are no prescription quotas in China. However, doctors can be fined for excessive or unreasonable prescriptions.

21.8.8 FINANCIAL INCENTIVES FOR PHYSICIANS

There are no financial incentives for physicians in order to control health care expenditure.

21.8.9 FINANCIAL INCENTIVES FOR PHARMACISTS

There are no financial incentives for pharmacists in order to control health care expenditure.

21.8.10 COPAYMENT FOR PATIENTS

Patients without health insurance must pay for all their health expenses. Patients with health insurance must pay a part of their health expenses, depending on their type of insurance. The Chinese system of copayment is rather complex and can vary from one insurance to another and from one province to another.

21.8.11 CHANGES IN THE PRICING MECHANISM

On June 1, 2015, China introduced new pricing regulations for drugs listed on the National Reimbursement Drug List or supplied by government programmes. "Government pricing" (fixed manufacturer prices) and "government-guided pricing" (fixed maximum retail price) were abolished for most drugs and replaced by new pricing mechanisms (Table 21.1).

For drugs with existing market competition, the new law introduced a "reimbursement standard" acting as a "guide for market prices" and understood as a form of internal reference pricing. Prices of the remaining drugs are to be established by multilateral negotiations involving pharmaceutical industry or by competitive tenders. However, many elements of the new pricing model remain unclear, for example, exact definition and methodology to calculate the "reimbursement standard" and framework for price negotiations. Supporting policies are missing, and coordination mechanisms have not been established yet.

TABLE 21.1

New Pricing Mechanisms Introduced on June 1, 2015

Group of Products	New Pricing Mechanism
Drugs included in the NRDL with existing market competition[a]	Reimbursement standard
Drugs included in the NRDL with little or no market competition (i.e., inpatient drugs and exclusively produced traditional Chinese medicines)[a]	Multilateral negotiations
Blood products and drugs procured and subsidized by the government (i.e., vaccines, HIV/AIDS drugs, and contraceptives)[a,b]	Tenders or multilateral negotiations

Source: National Development and Reform Commission. National Development and Reform Commission together with relevant departments announce the abolishment of government (guided) pricing for the majority of drugs, and push the drug pricing reform (5-5-2015). Available from: http://www.sdpc.gov.cn/xwzx/xwfb/201505/t20150505_690687.html.

Notes: NRDL: National Reimbursement Drug List

[a] Except Class I psychotropics and anesthetics, which remained under the "government-guided pricing"

[b] Applies to blood products not included in the NRDL

To inform the practical implementation of the new law, China is running, or plans to initiate, several pilot projects exercising applicability of various forms of internal and external reference pricing policies to the Chinese context. The city of Sanming is piloting an internal reference pricing for drugs with the same active ingredient and dosage form, and the "reimbursement standard" is set at the price of the cheapest generic from the group.[6] The municipality of Chongqing announced the initiation of the pilot project combining elements of internal reference pricing and a form of external reference pricing, namely comparison with prices of the same drug in other local areas (provinces and municipalities).[7,8] A number of cities are piloting an approach called the "2nd price negotiation," which allows hospitals to negotiate prices directly with suppliers.[9,10]

To conclude, pricing mechanism in China is changing, and it is still to be seen how the new law is implemented and executed in practice.

REFERENCES

1. The National Health and Family Planning Commission. Available from: http://www.nhfpc.gov.cn/ (accessed August 10, 2016).
2. Chine Government's official web Portal. Factfile: Ministry of Health. Available from: http://english.gov.cn/2005-10/09/content_75326.htm (accessed August 10, 2016).
3. The Ministry of Human Resources and Social Security. Available from: http://www.mohrss.gov.cn/ (accessed August 10, 2016).
4. The National Development and Reform Commission. Available from: http://en.ndrc.gov.cn/ (accessed August 10, 2016).
5. National Development and Reform Commission. National Development and Reform Commission together with relevant departments announce the abolishment of government (guided) pricing for the majority of drugs, and push the drug pricing reform (5-5-2015). Available from: http://www.sdpc.gov.cn/xwzx/xwfb/201505/t20150505_690687.html (accessed August 10, 2016).
6. Great wisdom Finance and economics (Shanghai). Sanming Healthcare reform: the reference price will set at the cheapest domestic generic to decrease drug prices. (10-11-2014). Available from: http://www.sm.gov.cn/ztzl/shyywstzgg/mtbd/201410/t20141011_274086.htm (accessed August 10, 2016).
7. Chongqing Municipal People's Government Office. Notification of the implementation of drugs payment standard for drugs within health insurance formulary (Trial) (3-23-2015). Available from: http://www.baiduyy.com/zbxx/zbxx_xx_30203.htm (accessed August 10, 2016).
8. Chongqing Municipal Human Resources and Social Security Bureau. Establishment of Chongqing drug price plan for drugs within the health insurance formulary (2014). Available from: http://www.vchale.com/ypzbxx_com/201799193_1_ce22cc8d14fd4f-0c9417f77826c30da9.html (accessed August 10, 2016).
9. State Council. State Council guidance on improving public hospitals centralized drug procurement (2-9-2015). Available from: http://www.gov.cn/zhengce/content/2015-02/28/content_9502.htm (accessed August 10, 2016).
10. Medicine Cloud Information. 1/4 of the cities might conduct the 2nd price negotiation, pharmaceutical companies entered the era of price negotiation (3-7-2015). Available from: http://www.yiyaojie.com/zb/zbzc/20150307/74192.html (accessed August 10, 2016).

Epilogue

With this impressive and unique book, Mondher Toumi makes a huge contribution to health care and health care policies. Market access in health care in fact means patient access to innovative (but often pricy) technologies. With health care budgets being limited, the challenge for policy makers is to spend health care money wisely. Therefore, price and reimbursement levels of medicines should correspond with an acceptable value for money from a societal perspective. The entire market access field is characterized by a plethora of factors that influence the final decision (to be or not to be ... reimbursed) and is subject to complex processes involving multiple stakeholders.

The author explains in clear terms decision criteria such as relative efficacy, relative effectiveness, cost-effectiveness, budget impact, and medical need. By reading this book, it becomes very clear to which extent and in which ways market access processes differ between countries. Moreover, and specific for the European setting, the delicate balance between competences at the EU level and its member states is tackled, but this book has a strong global focus with an in-depth discussion of market access aspects in, for instance, the United States, South Korea, China, and Japan.

Innovative ways of cooperation, involving multiple stakeholders, the interaction between regulators and payers, early dialogues, contracting, ... you name it, and it is in this book. Importantly, the text goes way beyond the usual buzz and hypes and aims to provide a neutral and correct view on all these aspects, founded on an enormous experience and knowledge.

Special attention is given to the fast-developing field of orphan medicines and to the very specific field of vaccines.

The problem with texts on market access is that this is a very rapidly evolving field, and, by the time the text appears, it risks already to be obsolete due to the rapid changing nature of the criteria and processes. Not so with this book. It is so rich and future oriented that it will easily become a standard work for many years to come.

What will the future bring? If we look at it from a societal perspective, there is a lot of progress to make. Article 25 of the Declaration of Human Rights states that every human being has the right to access to good medical care. Today we are far away from achieving this ambition. What can be done?

First and for all, there is a continuous need for improving the quality, effectiveness, and efficiency of our health systems. This means that there is an undisputable need for innovative health technologies, such as medicines, that help substantially reduce morbidity and mortality and improve quality of life. But the issue is that these so-called truly innovative technologies mostly come at an extra cost, and it is therefore of key importance to establish appropriate methods and procedures for pricing and reimbursement (P&R) of these technologies. Value should be rewarded, because without doing so, there will be no incentives to innovate and aim for added value. However, this reward cannot be infinite—we do not have all the money of the world to spend on health care. Societies should be more transparent, both to their citizens and to the innovative industries, about what those limits are. And it needs to be recognized that societies with a lower ability to pay should also pay less.

More consistency in value assessment, more transparency in funding processes and decisions, and fair prices that correctly reflect the value of innovations and account for the affordability from a societal perspective are required. Only then, can we succeed in guaranteeing both the sustainability of health care systems and the sustainability of innovation.

This book contains many ingredients and recipes that can help to get there. Let us start cooking.

Lieven Annemans
Professor of health economics, Ghent University, Belgium

Index

Note: Page numbers followed by 'f' and 't' refer to figures and tables, respectively.

For Product Safety Concerns and Information please contact our EU
representative GPSR@taylorandfrancis.com
Taylor & Francis Verlag GmbH, Kaufingerstraße 24, 80331 München, Germany

www.ingramcontent.com/pod-product-compliance
Ingram Content Group UK Ltd.
Pitfield, Milton Keynes, MK11 3LW, UK
UKHW011354240425
457818UK00023B/890